MW00587173

wildflowers *of*
the adirondacks

wildflowers *of* *the* adirondacks

DONALD J. LEOPOLD & LYTTON JOHN MUSSELMAN

JOHNS HOPKINS UNIVERSITY PRESS | BALTIMORE

Printed in China on acid-free paper
9 8 7 6 5 4 3 2 1

Johns Hopkins University Press
2715 North Charles Street
Baltimore, Maryland 21218-4363
www.press.jhu.edu

Library of Congress Cataloging-in-Publication Data

Names: Leopold, Donald Joseph, 1956– author. | Musselman, Lytton John, 1943– author.
Title: Wildflowers of the Adirondacks / Donald J. Leopold and Lytton John Musselman.
Description: Baltimore, Maryland : Johns Hopkins University Press, 2020. |
    Includes bibliographical references and index.
Identifiers: LCCN 2018051289 | ISBN 9781421431109 (pbk. : alk. paper) |
    ISBN 9781421431116 (electronic) | ISBN 1421431106 (pbk. : alk. paper) |
    ISBN 1421431114 (electronic)
Subjects: LCSH: Wild flowers—New York (State)—Adirondack Mountains—
    Identification. | Wild flowers—New York (State)—Identification.
Classification: LCC QK177 .L46 2019 | DDC 582.1309747—dc23
LC record available at https://lccn.loc.gov/2018051289

A catalog record for this book is available from the British Library.

*Special discounts are available for bulk purchases of this book. For more information,
please contact Special Sales at 410-516-6936 or specialsales@press.jhu.edu.*

Johns Hopkins University Press uses environmentally friendly book materials,
including recycled text paper that is composed of at least 30 percent post-consumer
waste, whenever possible.

Contents

Preface

The Adirondack Park, located solely within New York State, covers more than six million acres of land, with about 43% in forest preserve land and the rest in private land. The boundary of the Adirondack Mountains and Highlands roughly coincides with the legal boundaries of the park.

This is the largest protected area in the contiguous United States and widely regarded as a model for safeguarding a region's natural resources and biodiversity, whether on private or public lands. The park was established in 1894 and has doubled in size from the initial area of just over three million acres. Nationally, there might be more famous parklands in the United States—like Grand Canyon, Yellowstone, Yosemite, and Great Smoky Mountains National Parks—but these parks combined are only about 70% of the total size of the Adirondack Park.

The Adirondacks have an elevational range from 95 feet on the shores of Lake Champlain to 5,344 feet on Mount Marcy, more than ten thousand lakes, and thirty thousand miles of rivers and streams. This variety of conditions results in numerous natural communities, each defined by unique groups of plant and animal species. Broadly defined, these natural communities include extensive northern hardwood, mixed hardwood/conifer, and conifer (pine and spruce/fir) forests; marshes, bogs, fens, and shrub and forested swamps; and alpine zones. "Cultural" communities (that is, those resulting from human activity) occur as well, sometimes over significant areas like abandoned mine lands, old fields, and roadsides—each with its own array of native and non-native plant species.

Though a significant portion of the park is in a "forever wild" state, most of these lands are not without prior disturbance in the form of timber cutting, catastrophic fires, and major wind events. In addition, these many different kinds of

natural communities each exhibit a distinctive phase of ecological development, with early stages favoring some species;
later stages, others. And some natural communities are still
developing following the retreat of the glaciers that covered the
Adirondacks up to about ten thousand years ago. For example,
some of the large peatlands are still lacking substantial forest
canopy but presumably over time will be dominated by black
spruce and tamarack forests, unless climate change makes
these places much warmer, wetter, or drier. This mosaic of successional stages among so many natural communities results
in greater biodiversity, including that of wildflowers.

What a privilege it has been for us to have taught courses
each summer about the flora in the Adirondacks, through the
State University of New York College of Environmental Science
and Forestry's Cranberry Lake Biological Station. This book is
a result of these efforts, as well as our other teaching and research on the flora here and elsewhere over a combined ninety
years. Our students have taught us much over these decades,
including what specific information has helped them to learn
and get excited about these species. We hope that this book conveys this information and enthusiasm.

Acknowledgments

DJL thanks Michael Hough and Matt Young especially for their insights into the identification and ecology of the flora of New York. LJM acknowledges the Mary Payne Hogan Endowment at Old Dominion University for continued support of botanical scholarship. We thank Peter W. Schafran for field assistance and for the figures of *Epigaea repens* and *Sibbaldia tridentata*, which appear on pages 138 and 296, respectively; Lisa Wallace, Old Dominion University, for reviewing our treatment of *Platanthera*; and Harvey Ballard, Ohio University, for reviewing our treatment of *Viola*. We both also thank Beth Gianfagna for her superb copyediting.

How to Use This Book

This is a photographic guide to plants of the Adirondack Mountains and adjoining areas. Only herbaceous (non-woody) plants are included, except for a few woody species with very small stature and showy flowers. We have not included dichotomous keys, which have been described as an identification method devised by people who don't need the keys for people who can't use the keys. So, proper determination depends on identification based on photographs.

We have selected the images from a large database of our extensive work with these plants in the Adirondacks and elsewhere. Through teaching field courses for collectively more than seven decades, we have learned which distinguishing features are most helpful to students. The plants we have selected represent some of the most striking wildflowers, like orchids and goldenrods, as well as flowers easily overlooked, like cattails and hellebore, to give an overview of the wonderful diversity in the Adirondack flora.

Lastly, we have tried as much as possible to avoid botanical jargon.

Species Arrangement

Entries are arranged by predominant flower color. We consider this the most appropriate way to present the plants for a general audience. Using flower color does present difficulties, however, because some species have flowers of various colors. For instance, many species in the aster family have ray flowers (what appear to be the petals) that are quite differently colored from the prominent, center portion, or disk flowers, of the flower head. Some species (like goldthread and bunchberry) have plant parts that are petal-like but that are actually bracts surrounding less conspicuous flowers. We have grouped those species based on the predominant color, characterizing these two species as white, although the flowers are actually yellow. Some species change colors depending on the stage of flower development; for example, the white trillium usually becomes pink before the petals wither. Within each color group we include the image of the flower color most prevalent in that species and separately list the other colors that may be found in the same species.

We offer brief descriptions of each plant, aiming at an economy of words bereft of the tangle of descriptors used by botanists. We also include notes based on interesting features of the plant, its garden cultivation, and which plants are edible, as well as those that are toxic or likely to cause dermatitis. In most cases, these notes reflect our personal fascination with the plant.

Plant Names

Scientific names are taken from the New York Flora Atlas
(http://newyork.plantatlas.usf.edu). Likewise, common names
are from that atlas, though we have not always complied with
the punctuation used there. Some of the common names are
well known only locally. While we do not always agree with the
formal taxonomy of the New York Flora Atlas, for consistency,
all names of families and authors of scientific names are taken
from this site. Scientific names are usually of little interest to
beginners, but understanding nomenclature helps to clarify the
dynamic system of plant naming that is codified yet adaptable
to the most recent molecular studies. These names are also use-
ful when retrieving information on the plants.

Scientific Names Are Important

Two main components constitute the scientific name—the genus name and the specific epithet. Together these two parts make the single species name. The names or abbreviations that follow the species name are the authors of those names. This listing is an elegantly simple bibliography of plant names. For example, the scientific name of bunchberry is *Cornus canadensis* L., where *Cornus* is the genus, *canadensis* is the specific epithet, and the species name is *Cornus canadensis*. The "L." following the name is for Carl von Linné (usually Latinized to Linnaeus), which tells us that he was the first to describe that diminutive but beautiful plant, back in the 1700s. Sometimes the species' name will be followed by names or initials in parentheses, such as bastard toad flax, *Comandra umbellata* (L.) Nutt., signifying the plant was first described by Linnaeus but moved to a different genus by Thomas Nuttall (1786–1859). This documentation is essential for plant taxonomists, scientists who study the evolution and classification of plants.

wildflowers *of* *the* adirondacks

Introduction

What Are Wildflowers?

The term "wildflowers" has been defined both narrowly and broadly. In the narrowest sense, wildflowers are native, herbaceous species that have flowers that are showy enough to be readily noticed by the unaided eye. Spring wildflowers, like trilliums (*Trillium* species), are good examples. Wildflowers can also be defined as any uncultivated plant—that is, one that naturally occurs—with conspicuous flowers. Following this definition, woody plants that have attractive flowers and non-native plant species are included. However, for the purposes of this book, we include mostly herbaceous species that are native or naturalized in the Adirondacks and that have conspicuous flowers. In the species' descriptions we indicate which are native to the Adirondacks and which are introduced and have become naturalized. Some of these latter species, like purple loosestrife (*Lythrum salicaria*), although beautiful, are regarded as very serious invasive plants that threaten the biodiversity of the Adirondacks.

State-Protected Plants in the Adirondacks

Many of the plant species included in this guide are legally protected in New York State by the New York Protected Native Plants Regulations (6 NYCRR 193.3) and Environmental Conservation Law (Section 9-1503). These species are protected for at least one of two reasons: (1) the species is deemed "exploitably vulnerable" (EV), so it is protected to keep it from being harvested (because of its beauty or economic value) to the point of becoming rare; or (2) the species is truly rare. All of our native orchids, many species of the lily family (Liliaceae), all *Trillium* species, nearly all of our fern species, and other showy, native species are included in the EV category; some of these species are also rare.

There are three other New York State legal categories of plant protection besides exploitably vulnerable: endangered, threatened, and rare. Endangered species (E) are those with five or fewer extant (existing) sites; or, with fewer than one thousand individuals; or, that are restricted to fewer than four US Geological Survey (USGS) 7.5 minute topographic maps (which cover 7.5 minutes of latitude and longitude); or, that are federally listed as endangered. Threatened species (T) are those with six to fewer than twenty extant sites; or, with one thousand to fewer than three thousand individuals; or, that are restricted to not fewer than four or more than seven USGS 7.5 minute topographic maps; or, that are federally listed as threatened. Rare species (R) have twenty to thirty-five extant sites; or, three thousand to five thousand individuals statewide. As new locations for species are found and populations of other species disappear, the rarity status of some species will change. The most recent version of which species are listed in each category can be found on the website for the New York Natural Heritage Program.

Included in the listing of rarity status are New York Natural Heritage Program state ranks for the truly rare species of the Adirondacks. These categories, which also are noted in the species accounts in this book, are as follows:

S1 = Critically imperiled in New York State because of extreme rarity (five or fewer sites or very few remaining individuals) or extremely vulnerable to extirpation (elimination) from the state

S2 = Imperiled in the state because of rarity (six to twenty sites or few remaining individuals) or highly vulnerable to extirpation from the state

S3 = Rare in the state (usually twenty-one to one hundred extant sites)

SH = Historical, no extant sites known in the state

SX = Apparently extirpated from the state, with very low probability of rediscovery

Rarity is relative, in terms of time and location. Some species are rare by one measure but more abundant by another. Some New York State species are rare at the national level and hence are federally listed as threatened or endangered. None of the plant species that are currently known to occur in the Adirondacks are federally listed. However, it would not surprise some botanists if the small whorled pogonia [*Isotria medeoloides* (Pursh) Raf.*, nationally listed as endangered] is eventually found in Adirondack habitats, which seem similar to those where it is known to occur in New Hampshire.

Some plant species, even common ones, may be abundant during certain stages of forest succession and disappear at other stages. For example, populations of many of our native terrestrial orchids, such as the showy lady's slipper (*Cypripedium reginae*), dwindle as the forest canopy matures, allowing little light below. Other species, like the broad-lipped twayblade (*Neottia convallarioides*), seem to thrive under these deeply shaded conditions that result from forest maturation. Given that the Adirondack forests are still developing following extensive logging and burning in the late 1800s and early 1900s and after catastrophic wind and ice events since, locations and population sizes of our wildflower species should continue to be quite dynamic.

General Plant Community Types in the Adirondacks and Commonly Associated Wildflower Species

Different kinds of plant communities result from regional climate, geology, hydrology, and soils, as well as local site influences, such as natural and anthropogenic disturbances, slope aspect, and position. Types of communities can be defined using relatively coarse distinctions—for example, the few types recognized for all of New York State by Bray (1930)—or at much finer levels, as is done by the New York Natural Heritage Program, which recognizes 260 ecological communities in the state (Edinger et al. 2014). No community type is globally unique to the Adirondacks. For instance, although the alpine community is restricted to the High Peaks region and occupies less than one hundred acres, it is similar to the alpine communities in other northeastern states. The following summary combines many of the finer-scale community types recognized by the New York Natural Heritage Program into broader, more readily recognizable communities.

Open Water

These plant species are common in shallow, quiet waters of the Adirondacks. *Brasenia*, *Nuphar*, and *Nymphaea* grow in water up to a depth of about six and a half feet.

Brasenia schreberi	watershield
Eriocaulon aquaticum	northern pipewort
Lobelia dortmanna	water lobelia
Nuphar variegata	common yellow pond lily
Nymphaea odorata	fragrant white water lily
Pontederia cordata	pickerel weed
Potamogeton natans	floating-leaved pondweed

Utricularia cornuta	horned bladderwort
Utricularia purpurea	purple bladderwort

Muddy Flats and Sandy Shorelines

These plant species benefit from occasionally scarified (scraped by flowing water, ice, and wave action) substrates and little competition from adjacent plants.

Bidens cernua	nodding beggar ticks
Comarum palustre	marsh cinquefoil
Drosera intermedia	spatulate-leaved sundew
Eriocaulon aquaticum	northern pipewort
Euthamia graminifolia	flat-topped goldenrod
Gentiana linearis	narrow-leaved gentian
Gratiola aurea	golden hedge hyssop

Hypericum ellipticum	St. John's wort
Hypericum mutilum	dwarf St. John's wort
Hypericum virginicum	Virginia marsh St. John's wort
Lobelia cardinalis	cardinal flower
Lobelia dortmanna	water lobelia
Lycopus americanus	American bugleweed
Lycopus uniflorus	northern bugleweed
Lysimachia terrestris	swamp candles
Mimulus ringens	Allegheny monkey flower
Persicaria amphibia	American water smartweed
Pontederia cordata	pickerel weed
Sagittaria graminea	grass-leaved arrowhead
Sagittaria latifolia	arrowhead
Scutellaria galericulata	marsh skullcap
Scutellaria lateriflora	mad-dog skullcap
Sium suave	hemlock water parsnip
Sparganium americanum	bur-reed
Utricularia resupinata	reclined bladderwort

Deep Emergent Marshes

Besides the wildflower species listed here, deep emergent marshes are often dominated by cattail taxa (the native *Typha latifolia*, the introduced *T. angustifolia*, and the hybrid between these two species, *T. × glauca*); the introduced and highly invasive purple loosestrife (*Lythrum salicaria*); and the common reed (*Phragmites communis*). Some robust, grasslike plants such as bulrushes (*Schoenoplectus* spp.) are also common. These wetlands often grade into open water.

Comarum palustre	marsh cinquefoil
Decodon verticillatus	water willow
Lobelia siphilitica	great blue lobelia
Nasturtium officinale	water cress
Pontederia cordata	pickerel weed
Sagittaria graminea	grass-leaved arrowhead
Sagittaria latifolia	arrowhead
Sium suave	hemlock water parsnip
Typha angustifolia	narrow-leaved cattail
Typha latifolia	wide-leaved cattail
Typha × glauca	hybrid cattail
Utricularia vulgaris	greater bladderwort
Veronica americana	American brooklime

Shallow Emergent Marshes

These wetlands have a relatively high diversity of plant species unless they are dominated by cattail taxa (*Typha latifolia*, *T. angustifolia*, and *T.×glauca*); the introduced and highly invasive purple loosestrife (*Lythrum salicaria*); and the European common reed [*Phragmites australis* (Cav.) Trin. ex Steud.].

Besides the species with showy flowers that are listed here, many grasslike species belonging to the genus *Carex* (sedge) are abundant. Shallow emergent marshes often are adjacent to or intermixed with deep emergent marshes and shrub-dominated wetlands. Because marshes are generally more hydrologically dynamic than bogs and fens, including having longer periods of drawdown or dry conditions, they do not accumulate the large amounts of organic matter that defines those peatlands.

Angelica atropurpurea	purple-stemmed angelica
Asclepias incarnata	swamp milkweed
Bidens cernua	nodding beggar ticks
Bidens frondosa	devil's beggar ticks
Caltha palustris	marsh marigold
Campanula aparinoides	marsh bellflower
Chelone glabra	turtlehead
Cicuta bulbifera	bulb-bearing water hemlock
Cicuta maculata	spotted water hemlock
Comarum palustris	marsh cinquefoil
Doellingeria umbellata	flat-topped aster
Eupatorium perfoliatum	boneset
Euthamia graminifolia	flat-topped goldenrod
Eutrochium fistulosum	hollow Joe Pye weed
Eutrochium maculatum	spotted Joe Pye weed
Eutrochium purpureum	purple-node Joe Pye weed
Helenium autumnale	sneezeweed
Hypericum virginicum	Virginia marsh St. John's wort
Impatiens capensis	spotted jewelweed
Iris versicolor	blue flag iris
Lobelia cardinalis	cardinal flower
Lobelia siphilitica	great blue lobelia
Lycopus americanus	American bugleweed
Lycopus uniflorus	northern bugleweed
Lysimachia terrestris	swamp candles

Lysimachia thyrsiflora	tufted loosestrife
Lythrum salicaria	purple loosestrife
Sagittaria graminea	grass-leaved arrowhead
Sagittaria latifolia	arrowhead
Scutellaria galericulata	marsh skullcap
Sium suave	hemlock water parsnip
Solidago rugosa	wrinkle-leaved goldenrod
Sparganium americanum	bur-reed
Symphyotrichum prenanthoides	crooked-stemmed aster
Symphyotrichum puniceum	purple-stemmed aster
Thalictrum pubescens	tall meadow rue
Verbena hastata	blue vervain

Moist Meadows

Soils of moist meadows are not saturated long enough to develop the characteristics and related plant stresses of wetland soils, but are sufficiently wet to support plants that do not thrive on drier substrates. Many moist meadows throughout the Adirondacks that developed following beaver flooding and subsequent dam abandonment are dominated by sedges and other grasslike plants.

Angelica atropurpurea	purple-stemmed angelica
Barbarea vulgaris	yellow rocket
Cirsium muticum	swamp thistle
Clematis virginiana	virgin's bower

Euthamia graminifolia	flat-topped goldenrod
Gentiana linearis	narrow-leaved gentian
Hesperis matronalis	dame's rocket
Houstonia caerulea	common bluets
Lilium canadense	Canada lily
Lythrum salicaria	purple loosestrife
Mentha arvensis	field mint
Pastinaca sativa	wild parsnip
Platanthera lacera	ragged fringed orchid
Prunella vulgaris	self-heal, heal-all
Ranunculus acris	common buttercup
Rudbeckia laciniata	green-headed coneflower
Sisyrinchium angustifolium	narrow-leaved blue-eyed grass
Thalictrum pubescens	tall meadow rue
Viola sororia	common violet
Zizia aurea	golden Alexanders

Streambanks

Some of the species that grow along streambanks benefit from the occasional bank scouring and bare substrate made available for seed germination. Speckled alder (*Alnus incana* ssp. *rugosa*) often forms dense thickets along streambanks, especially smaller streams with slower water.

Angelica atropurpurea	purple-stemmed angelica
Apios americana	groundnut
Apocynum cannabinum	Indian hemp
Arisaema triphyllum	Jack-in-the-pulpit
Cuscuta gronovii	common dodder
Doellingeria umbellata	flat-topped aster

Eutrochium maculatum	spotted Joe Pye weed
Hypericum perforatum	common St. John's wort
Impatiens capensis	spotted jewelweed
Lobelia cardinalis	cardinal flower
Lobelia siphilitica	great blue lobelia
Lythrum salicaria	purple loosestrife
Melilotus albus	white sweet clover
Mimulus ringens	Allegheny monkey flower
Neottia auriculata	auricled twayblade
Orobanche uniflora	one-flowered broomrape
Persicaria sagittata	arrow-leaved tearthumb
Solidago rugosa	wrinkle-leaved goldenrod
Symphyotrichum puniceum	purple-stemmed aster
Thalictrum pubescens	tall meadow rue
Veronica americana	American brooklime
Viola cucullata	marsh blue violet

Floodplain Forests

Adirondack floodplain forests are dominated by silver maple
(*Acer saccharinum* L.), American elm (*Ulmus americana* L.),
green ash (*Fraxinus pennsylvanica* Marshal), Freeman maple (*Acer saccharinum* × *A. rubrum* A. E. Murray), boxelder
(*Acer negundo* L.), and black ash (*Fraxinus nigra* Marshall).
American hornbeam (*Carpinus caroliniana* Walter) and red
maple (*Acer rubrum* L.) can also be common. Trees more typical
of Adirondack uplands, like yellow birch (*Betula alleghaniensis*
Britton), occur on elevated terrain within floodplains.

Ageratina altissima	white snakeroot
Amphicarpaea bracteata	hog peanut
Angelica atropurpurea	purple-stemmed angelica
Apios americana	groundnut
Arisaema triphyllum	Jack-in-the-pulpit
Bidens frondosa	devil's beggar ticks
Cirsium muticum	swamp thistle
Hesperis matronalis	dame's rocket
Impatiens capensis	spotted jewelweed
Impatiens pallida	pale jewelweed
Laportea canadensis	wood nettle
Lycopus uniflorus	northern bugleweed
Lysimachia terrestris	swamp candles
Scutellaria lateriflora	mad-dog skullcap

Solidago canadensis	Canada goldenrod
Solidago gigantea	smooth goldenrod
Symplocarpus foetidus	skunk cabbage
Thalictrum pubescens	tall meadow rue
Urtica dioica	European stinging nettle

Lowland Hardwood Forests

These wetland forests develop in poorly drained depressions within the landscape and are generally dominated by red maple, black ash, tamarack or eastern larch [*Larix laricin*a (Du Roi) K. Koch], eastern hemlock, and yellow birch. Many other tree species can also occur, depending on the extent and duration of water above or near the ground surface. These forests are often located further away from floodplain forests or are entirely disconnected from them. Water and substrate pH, as influenced by calcium, have significant effects on plant species composition. Understory composition, including wildflower diversity, also depends greatly on the degree of canopy openness.

Aralia nudicaulis	wild sarsaparilla
Bidens cernua	nodding beggar ticks
Bidens frondosa	devil's beggar ticks
Calla palustris	wild calla
Caltha palustris	marsh marigold
Chelone glabra	turtlehead
Clintonia borealis	blue bead lily
Coptis trifolia	goldthread

Cornus canadensis	bunchberry
Impatiens capensis	spotted jewelweed
Lycopus americanus	American bugleweed
Lycopus uniflorus	northern bugleweed
Lysimachia borealis	starflower
Lysimachia thyrsiflora	tufted loosestrife
Maianthemum canadense	Canada mayflower
Mitchella repens	partridge berry
Solidago patula	rough-leaved goldenrod
Thalictrum pubescens	tall meadow rue

Open, Acidic Peatlands (Bogs)

Open canopy, acidic peatlands, or bogs, are readily recognized by their carpets of *Sphagnum* moss species; black spruce [*Picea mariana* (Mill.) Britton, Sterns & Poggenb.] and tamarack or eastern larch; numerous shrub species in the heath family

(Ericaceae), including the trailing woody species known as cranberries (*Vaccinium macrocarpon* Aiton and *V. oxycoccus* L.); and insectivorous species, especially purple pitcherplant (*Sarracenia purpurea*). These wetlands usually occur in low positions in the landscape. The water in bogs is strongly acidic, with a pH of about 3.8 to 5.5, so this type of plant community also includes wetlands known as "poor" fens—those that receive little calcium-rich groundwater (see "Open, Alkaline Peatlands" below).

Adirondack peatlands are among the most interesting natural communities of all in New York, in large part because they support so many terrestrial orchid and insectivorous species. Although not quite as easy to explore as Adirondack uplands, anyone really interested in seeing some of our rarest and most beautiful native plant species will endure the challenges of searching these soggy areas for wildflowers.

Arethusa bulbosa	dragon's mouth
Calla palustris	wild calla
Calopogon tuberosus	grass pink
Drosera intermedia	spatulate-leaved sundew
Drosera rotundifolia	round-leaved sundew
Maianthemum trifolium	three-leaved Solomon's seal
Menyanthes trifoliata	buckbean
Oclemena nemoralis	bog aster
Platanthera blephariglottis	white fringed orchid
Pogonia ophioglossoides	rose pogonia
Sarracenia purpurea	purple pitcherplant
Utricularia cornuta	horned bladderwort

Black Spruce Tamarack Forests

Open bogs, without hydrologic or other disturbances, eventually succeed to a black spruce (*Picea mariana*) and tamarack (*Larix laricina*) forest. Many of the showy wildflowers of open bogs disappear in these densely shaded forests, especially the characteristic orchids. Eventually, many of the woody heath family species disappear owing to the dense shade cast by the dominant conifers. The pitcherplant survives for decades after forest canopy closure, but the pitchers become flatter and longer in length and are green instead of red and purple.

Coptis trifolia	goldthread
Cornus canadensis	bunchberry
Gaultheria hispidula	snowberry
Sarracenia purpurea	purple pitcherplant

Open, Alkaline Peatlands (Fens)

Open canopy, near-neutral to alkaline (5.5 to 7.5 pH) wet-lands—typically of organic soils, although the substrate can be marl (basically, calcium carbonate deposits)—occur in depressions or near breaks in slopes where bedrock or glacial material is high in calcium and the groundwater that is discharged and that creates these wetlands is also high in this element. Limy habitats, either in lowlands or uplands, are relatively scarce in the Adirondacks, hence some of the rarest plant species in these mountains are found in these natural communities.

Fens are differentiated in terms of several key factors: water level, successional stage, and vegetational composition, which is the result of the chemistry of the groundwater discharged to the wetland. The greater the calcium and other minerals discharged, the "richer" the fen. Fens that have so little mineral input that they resemble bogs are known as "poor fens." Some fens are dominated by herbaceous species and others by shrub species like "bog" birch (*Betula pumila* L.) and hoary willow (*Salix candida* Flüggé ex Willd.). The end point of fen succession is northern white-cedar swamp, which is really a much more diverse assemblage of trees, shrubs, and herbaceous species than the common name suggests.

Arethusa bulbosa	dragon's mouth
Asclepias incarnata	swamp milkweed
Campanula aparinoides	marsh bellflower
Cirsium muticum	swamp thistle
Cypripedium reginae	showy lady's slipper
Geum rivale	purple avens
Iris versicolor	blue flag iris
Liparis loeselii	fen twayblade
Lobelia kalmii	Kalm's lobelia

Maianthemum stellatum	starry Solomon's seal
Menyanthes trifoliata	buckbean
Packera aurea	golden ragwort
Parnassia glauca	grass-of-Parnassus
Platanthera aquilonis	northern green bog orchid
Platanthera dilatata	tall white bog orchid
Platanthera grandiflora	large purple fringed orchid
Platanthera huronensis	Lake Huron bog orchid
Platanthera lacera	ragged fringed orchid
Platanthera psycodes	lesser purple fringed orchid
Pyrola asarifolia	pink wintergreen
Rudbeckia laciniata	green-headed coneflower
Solidago patula	rough-leaved goldenrod
Solidago uliginosa	bog goldenrod
Spiranthes cernua	nodding ladies' tresses
Spiranthes lucida	shining ladies' tresses
Spiranthes romanzoffiana	hooded ladies' tresses
Symplocarpus foetidus	skunk cabbage
Thalictrum pubescens	tall meadow rue

Cedar Swamps

Over time and lacking disturbance, open canopy, alkaline peatlands develop a closed canopy forest dominated by northern white-cedar (*Thuja occidentalis* L.) and other conifers, including balsam fir [*Abies balsamea* (L.) Mill.], tamarack or eastern larch (*Larix laricina*), eastern white pine (*Pinus strobus* L.), eastern hemlock [*Tsuga canadensis* (L.) Carrière],

and white spruce [*Picea glauca* (Moench) Voss], where it occurs in the Adirondacks. The occurrence of tamarack, and
even black spruce (*Picea mariana*) in these wetlands is common, indicating that these characteristic species of acidic
bogs do not require those low pH conditions to thrive. Black
ash (*Fraxinus nigra*) is a widespread deciduous tree in these
wetlands, as is the otherwise upland-occurring yellow birch
(*Betula alleghaniensis*).

Calcium-enriched groundwater discharge is usually associated with these wetlands, which also support a large number
of upland forest species because of the extreme microtopography, especially naturally occurring mounds (hummocks) that
are totally removed from any groundwater influences. Because
of the great variation between wet and dry conditions and high
and low pH in these microsites, cedar swamp wetlands are quite
diverse and support numerous species of terrestrial orchids.

Caltha palustris	marsh marigold
Clintonia borealis	blue bead lily
Coptis trifolia	goldthread
Corallorhiza trifida	early coralroot
Cornus canadensis	bunchberry
Cypripedium arietinum	ram's head lady's slipper
Cypripedium parviflorum	yellow lady's slipper
Cypripedium reginae	showy lady's slipper
Gaultheria hispidula	snowberry
Geum rivale	purple avens
Impatiens capensis	spotted jewelweed
Linnaea borealis	twinflower
Lysimachia borealis	starflower
Maianthemum canadense	Canada mayflower
Malaxis monophyllos var. *brachypoda*	white adder's mouth
Malaxis unifolia	green adder's mouth
Mitchella repens	partridge berry
Mitella nuda	naked mitrewort
Moneses uniflora	one-flowered wintergreen
Neottia convallarioides	broad-lipped twayblade
Oxalis montana	northern wood sorrel
Packera aurea	golden ragwort
Platanthera flava var. *herbiola*	northern tubercled orchid
Platanthera grandiflora	large purple fringed orchid
Platanthera obtusata	blunt-leaved orchid
Platanthera psycodes	lesser purple fringed orchid

Pyrola asarifolia	pink wintergreen
Rubus repens	dewdrop
Symphyotrichum boreale	northern bog aster
Tiarella cordifolia	foamflower
Viola cucullata	marsh blue violet

Lowland Conifer Forests

Red spruce (*Picea rubens*), balsam fir (*Abies balsamea*), black spruce (*Picea mariana*), and white spruce (*Picea glauca*) can dominate these forested wetlands. Yellow birch (*Betula alleghaniensis*), eastern white pine (*Pinus strobus*), black ash (*Fraxinus nigra*), tamarack (*Larix laricina*), northern white-cedar (*Thuja occidentalis*), and eastern hemlock (*Tsuga canadensis*) often occur in fewer numbers. Many of the wildflowers in this community are also common in upland forests but are found in these lowlands because of the pronounced microtopography, especially raised microsites like old tree stumps, windthrows, and downed logs.

Clintonia borealis	blue bead lily
Coptis trifolia	goldthread
Cornus canadensis	bunchberry
Cypripedium acaule	pink lady's slipper
Gaultheria hispidula	snowberry
Maianthemum canadense	Canada mayflower
Maianthemum trifolium	three-leaved Solomon's seal
Moneses uniflora	one-flowered wintergreen

Oclemena acuminata	whorled wood aster
Orthilia secunda	one-sided wintergreen
Linnaea borealis	twinflower
Lysimachia borealis	starflower
Neottia cordata	heart-leaved twayblade
Oxalis montana	northern wood sorrel
Pyrola asarifolia	pink wintergreen
Rubus repens	dewdrop

Mixed Hardwood Forests

These upland forests, when dominated by sugar maple (*Acer saccharum*), American beech (*Fagus grandifolia*), and yellow birch (*Betula alleghaniensis*), are known as the northern hardwood forest. Red maple (*Acer rubrum*) and black cherry (*Prunus serotina* Ehrh.) are often common in these forests as well. On richer sites that are influenced by calcium bedrock or glacial till, American basswood (*Tilia americana* L.) and white ash (*Fraxinus americana*) are more prominent, and wildflower species' richness increases.

Actaea pachypoda	white baneberry
Actaea rubra	red baneberry
Ageratina altissima	white snakeroot
Allium tricoccum	wild leek
Anemone quinquefolia	wood anemone
Aquilegia canadensis	wild columbine
Aralia nudicaulis	wild sarsaparilla
Aralia racemosa	spikenard
Arisaema triphyllum	Jack-in-the-pulpit
Asarum canadense	wild ginger

Cardamine concatenata	cut-leaved toothwort
Cardamine diphylla	two-leaved toothwort
Caulophyllum thalictroides	blue cohosh
Circaea alpina	dwarf enchanter's nightshade
Claytonia caroliniana	Carolina spring beauty
Claytonia virginica	eastern spring beauty
Cypripedium parviflorum	yellow lady's slipper
Desmodium canadense	showy tick trefoil
Dicentra canadensis	squirrel corn
Dicentra cucullaria	Dutchman's breeches
Epifagus virginiana	beechdrops
Erythronium americanum	yellow trout lily
Eurybia divaricata	white wood aster
Eurybia macrophylla	large-leaved aster
Galearis spectabilis	showy orchid
Geranium maculatum	wild geranium
Geranium robertianum	herb Robert
Geum fragarioides	barren strawberry
Hepatica acutiloba	sharp-lobed hepatica
Hepatica americana	round-lobed hepatica
Hydrophyllum virginianum	waterleaf
Lactuca biennis	tall blue lettuce
Lactuca canadensis	tall lettuce
Laportea canadensis	wood nettle
Lysimachia borealis	starflower
Maianthemum canadense	Canada mayflower
Maianthemum racemosum	false Solomon's seal

Medeola virginiana	Indian cucumber root
Micranthes virginiensis	early saxifrage
Mitchella repens	partridge berry
Mitella diphylla	two-leaved mitrewort
Monotropa uniflora	Indian pipe
Nabalus albus	white rattlesnake root
Nabalus altissimus	tall rattlesnake root
Oclemena acuminata	whorled wood aster
Osmorhiza claytonii	bland sweet cicely
Oxalis montana	northern wood sorrel
Panax quinquefolius	American ginseng
Panax trifolius	dwarf ginseng
Pedicularis canadensis	wood betony
Platanthera hookeri	Hooker's orchid
Platanthera orbiculata	small-spurred round-leaved orchid
Podophyllum peltatum	mayapple
Polygonatum pubescens	hairy Solomon's seal
Sanguinaria canadensis	bloodroot
Sanicula marilandica	Maryland sanicle
Solidago caesia	blue-stemmed goldenrod
Solidago flexicaulis	zigzag goldenrod
Streptopus lanceolatus	rose twisted stalk
Symphyotrichum cordifolium	heart-leaved aster
Thalictrum dioicum	early meadow rue
Tiarella cordifolia	foamflower
Trillium erectum	purple trillium

Trillium undulatum	painted trillium
Uvularia sessilifolia	wild oats
Viola canadensis	Canada violet
Viola pubescens	downy yellow-stemmed violet
Viola renifolia	kidney-leaved violet
Viola rostrata	long-spurred violet
Viola rotundifolia	early yellow violet

Mixed Hardwood Conifer Forests

These upland forests are dominated primarily by sugar maple (*Acer saccharum*), American beech (*Fagus grandifolia* Ehrh.), and yellow birch (*Betula alleghaniensis*), but they can include a significant component of eastern hemlock (*Tsuga canadensis*). Red maple (*Acer rubrum*), red spruce (*Picea rubens*), and eastern white pine (*Pinus strobus*) can also be common. In many places, red spruce was, historically, likely the most abundant species, but heavy harvesting of this species in the late 1800s and early 1900s, as well as catastrophic fires in the early 1900s, greatly changed the composition of Adirondack forests.

Aralia nudicaulis	wild sarsaparilla
Arisaema triphyllum	Jack-in-the-pulpit
Epifagus virginiana	beechdrops
Eurybia divaricata	white wood aster
Goodyera tesselata	checkered rattlesnake plantain
Lysimachia borealis	starflower

Maianthemum canadense	Canada mayflower
Medeola virginiana	Indian cucumber root
Mitchella repens	partridge berry
Monotropa uniflora	Indian pipe
Oxalis montana	northern wood sorrel
Rubus repens	dewdrop
Streptopus lanceolatus	rose twisted stalk
Tiarella cordifolia	foamflower
Trillium erectum	purple trillium
Uvularia sessilifolia	wild oats
Viola rotundifolia	early yellow violet

Mixed Conifer Hardwood Forests

Red spruce (*Picea rubens*), eastern white pine (*Pinus strobus*), or eastern hemlock (*Tsuga canadensis*) are significant

components of these upland forests, joined by sugar maple (*Acer saccharum*), yellow birch (*Betula alleghaniensis*),
American beech (*Fagus grandifolia*), and red maple (*Acer rubrum*). Balsam fir (*Abies balsamea*) is usually a minor component. Striped maple (*Acer pensylvanicum* L.) and mountain
maple (*Acer spicatum* Lam.) are in the subcanopy.

Aralia nudicaulis	wild sarsaparilla
Clintonia borealis	blue bead lily
Coptis trifolia	goldthread
Cornus canadensis	bunchberry
Epigaea repens	trailing arbutus
Eurybia macrophylla	large-leaved aster
Fragaria virginiana	common wild strawberry
Gaultheria procumbens	teaberry
Linnaea borealis	twinflower
Lysimachia borealis	starflower
Maianthemum canadense	Canada mayflower
Medeola virginiana	Indian cucumber root
Mitella diphylla	two-leaved mitrewort
Oclemena acuminata	whorled wood aster
Oxalis montana	northern wood sorrel
Platanthera hookeri	Hooker's orchid
Polygaloides paucifolia	gaywings
Streptopus lanceolatus	rose twisted stalk
Trillium undulatum	painted trillium
Viola blanda	sweet white violet

Pine and Hemlock Forests

Eastern white pine (*Pinus strobus*), red pine (*Pinus resinosa* Aiton), and eastern hemlock (*Tsuga canadensis*) dominate this upland forest type, typically on coarse, very well-drained soils. Paper birch (*Betula papyrifera* Marshall) can occur as remnants in places where the dominants earlier colonized

burned sites. Site conditions are acidic, dry, and deeply shaded, to which relatively few herbaceous species are well adapted.

Apocynum androsaemifolium	spreading dogbane
Aralia nudicaulis	wild sarsaparilla
Chimaphila umbellata	pipsissewa
Clintonia borealis	blue bead lily
Coptis trifolia	goldthread
Cornus canadensis	bunchberry
Cypripedium acaule	pink lady's slipper
Epigaea repens	trailing arbutus
Gaultheria procumbens	teaberry
Geum fragarioides	barren strawberry
Goodyera repens	dwarf rattlesnake plantain
Goodyera tesselata	checkered rattlesnake plantain

Linnaea borealis	twinflower
Lysimachia borealis	starflower
Maianthemum canadense	Canada mayflower
Melampyrum lineare	cow-wheat
Mitchella repens	partridge berry
Monotropa uniflora	Indian pipe
Orthilia secunda	one-sided wintergreen
Oxalis montana	northern wood sorrel
Pyrola americana	round-leaved shinleaf
Pyrola elliptica	common shinleaf
Solidago hispida	hairy goldenrod
Trillium undulatum	painted trillium
Viola labradorica	American dog violet

Mountain Spruce Fir Forests

Red spruce (*Picea rubens*) and balsam fir (*Abies balsamea*) are co-dominants of this upland community at higher elevations in the Adirondacks. Mountain paper birch (*Betula cordifolia* Regel), yellow birch (*Betula alleghaniensis*), and American mountain-ash (*Sorbus americana* Marshall) are less abundant.

Aralia nudicaulis	wild sarsaparilla
Clintonia borealis	blue bead lily
Coptis trifolia	goldthread
Cornus canadensis	bunchberry
Maianthemum canadense	Canada mayflower
Neottia cordata	heart-leaved twayblade

Oclemena acuminata	whorled wood aster
Oxalis montana	northern wood sorrel
Solidago macrophylla	large-leaved goldenrod

Mountain Fir Forests

Balsam fir (*Abies balsamea*) dominates the canopy of these forests at high elevations up to the alpine community. Mountain paper birch (*Betula cordifolia*), red spruce (*Picea rubens*), and American mountain-ash (*Sorbus americana*) occur as lesser components.

Clintonia borealis	blue bead lily
Coptis trifolia	goldthread
Cornus canadensis	bunchberry
Erythronium americanum	yellow trout lily
Oclemena acuminata	whorled wood aster

Oxalis montana	northern wood sorrel
Solidago macrophylla	large-leaved goldenrod
Streptopus amplexifolius	clasping-leaved twisted stalk

Alpine

Many woody and graminoid (grasslike) species dominate the one hundred or so acres of alpine communities in the Adirondacks. Stunted balsam fir (*Abies balsamea*), red spruce (*Picea rubens*), and paper birch (*Betula papyrifera*) are often abundant, along with numerous shrub species that are also common in acidic peatlands, as well as species found in no other natural community in New York.

Clintonia borealis	blue bead lily
Coptis trifolia	goldthread
Cornus canadensis	bunchberry

Lysimachia borealis	starflower
Maianthemum canadense	Canada mayflower
Mononeuria groenlandica	mountain sandwort
Nabalus boottii	Boott's rattlesnake root
Nabalus trifoliolatus	three-leaved rattlesnake root
Sibbaldia tridentata	three-toothed cinquefoil
Solidago cutleri	Cutler's alpine goldenrod
Solidago macrophylla	large-leaved goldenrod
Veratrum viride	false hellebore

Old Fields and Waste Places

The plant species composition of old fields and waste places varies significantly based on the management history of the site (including time since abandonment), soil moisture and fertility, and available light. Old fields are often dominated by species in the aster family, especially asters and goldenrods. The state-protected butterfly weed (*Asclepias tuberosa*) can be especially common on these sites, as well as along roadsides, if the conditions are dry and sunny.

While many species in these habitats have been introduced from Europe and Asia and are now naturalized in the Adirondacks and far beyond (see the entries for individual species for native status), most of them are important nectar plants for butterflies and moths, and provide other important ecological functions under conditions in which relatively few species can thrive.

Achillea millefolium	yarrow
Anaphalis margaritacea	pearly everlasting
Antennaria neglecta	field pussytoes
Asclepias syriaca	common milkweed
Asclepias tuberosa	butterfly weed
Campanula rotundifolia	harebell
Centaurea jacea	brown knapweed
Centaurea nigra	black knapweed
Centaurea stoebe ssp. *micranthos*	spotted knapweed
Cichorium intybus	chicory
Cirsium arvense	Canada thistle
Cirsium vulgare	bull thistle
Daucus carota	wild carrot

Dipsacus fullonum	teasel
Echium vulgare	viper's bugloss
Erigeron philadelphicus	Philadelphia fleabane
Euphrasia stricta	upright eyebright
Fragaria virginiana	common wild strawberry
Hemerocallis fulva	day lily
Hypericum perforatum	common St. John's wort
Inula helenium	elecampane
Leucanthemum vulgare	oxeye daisy
Linaria vulgaris	butter-and-eggs
Lobelia inflata	Indian tobacco
Lobelia spicata	common spiked lobelia
Malva moschata	musk mallow
Melilotus albus	white sweet clover
Melilotus officinalis	yellow sweet clover
Mentha canadensis	American wild mint
Monarda fistulosa	wild bergamot
Oenothera biennis	evening primrose
Pastinaca sativa	wild parsnip
Pilosella aurantiaca	orange hawkweed
Pilosella caespitosa	yellow hawkweed
Pilosella piloselloides	glaucous-leaved hawkweed
Potentilla norvegica	ternate-leaved cinquefoil
Potentilla recta	sulphur cinquefoil
Prunella vulgaris	self-heal, heal-all
Rudbeckia hirta	black-eyed Susan
Rumex acetosella	sheep sorrel

Saponaria officinalis	bouncing bet
Silene latifolia	white campion
Silene vulgaris	bladder campion
Solidago canadensis	Canada goldenrod
Solidago juncea	early goldenrod
Solidago nemoralis	gray goldenrod
Solidago rugosa	wrinkle-leaved goldenrod
Symphyotrichum lateriflorum	calico aster
Symphyotrichum novae-angliae	New England aster
Symphyotrichum novi-belgii	New York aster
Symphyotrichum pilosum	frostweed aster
Tragopogon pratensis	meadow salsify, goat's beard
Tussilago farfara	coltsfoot
Urtica dioica	European stinging nettle
Verbascum thapsus	common mullein
Veronica officinalis	common speedwell
Veronica serpyllifolia	common thyme-leaved speedwell
Viola sororia	common violet

Roadsides and Adjacent Ditches

The plant species found along roadsides are determined by the light and substrate conditions (including type of substrate and its moisture levels), roadside management (such as the mowing schedule; herbicide type, if applied; and grading), and type of vegetation adjacent to the roadside. For example, forest understory species that thrive in high light, like the white trillium

(*Trillium grandiflorum*), often form dense bands along road-sides when other conditions are also ideal. While traveling to many sites over the past few decades to show students exemplary plant communities and their plant species, we have spent much time stopping along the way to explore roadsides, including ditches (indicated below by the letter *D*).

Anaphalis margaritacea	pearly everlasting
Anemone canadensis	Canadian anemone
Apocynum androsaemifolium	spreading dogbane
Apocynum cannabinum	Indian hemp
Asclepias incarnata (D)	swamp milkweed
Asclepias tuberosa	butterfly weed
Bidens cernua (D)	nodding beggar ticks
Bidens frondosa (D)	devil's beggar ticks
Chamaenerion angustifolium	fireweed
Chelone glabra (D)	turtlehead

Cypripedium reginae (D)	showy lady's slipper
Gentiana linearis (D)	narrow-leaved gentian
Hesperis matronalis	dame's rocket
Inula helenium	elecampane
Iris versicolor (D)	blue flag iris
Liparis loeselii (D)	fen twayblade
Lobelia cardinalis (D)	cardinal flower
Lobelia spicata (D)	common spiked lobelia
Lythrum salicaria (D)	purple loosestrife
Maianthemum racemosum	false Solomon's seal
Maianthemum stellatum	starry Solomon's seal
Nasturtium officinale	water cress
Platanthera aquilonis (D)	northern green bog orchid
Platanthera clavellata (D)	little club-spur orchid
Platanthera dilatata (D)	tall white bog orchid
Platanthera grandiflora (D)	large purple fringed orchid
Platanthera huronensis (D)	Lake Huron bog orchid
Platanthera lacera (D)	ragged fringed orchid
Polygonatum pubescens	hairy Solomon's seal
Saponaria officinalis	bouncing bet
Spiranthes cernua (D)	nodding ladies' tresses
Spiranthes lucida (D)	shining ladies' tresses
Symphyotrichum puniceum (D)	purple-stemmed aster
Trillium grandiflorum	white trillium
Tussilago farfara	coltsfoot
Veronica americana (D)	American brooklime
Viola labradorica	American dog violet

Overview of Special Groups of Adirondack Wildflowers

Several subsets of Adirondack wildflowers belong to particularly showy and diverse families and genera. We include this material here because of its importance and because it does not tidily fit into sections of flower colors, given the range of flower colors within genera and families. Overviews in this section are to the aster and orchid families (Asteraceae and Orchidaceae) and the *Trillium* and *Viola* species.

Aster or Composite Family (Asteraceae)

The blooms of the aster or composite family are actually composed of one or two types of different flowers within a compact head: namely, ray flowers and disk flowers. Ray flowers are straplike and typically showy (they are absent in *Anaphalis*, *Antennaria*, *Centaurea*, *Cirsium*, *Eupatorium*, *Eutrochium*). Disk flowers are tiny and tubelike, and make up the center of the head when they occur (they are absent in *Cichorium, Hieracium, Lactuca, Pilosella, Nabalus,* and *Tragopogon*). The leaves and stems of these latter six genera contain a milky sap. The nectar and pollen of flowers in this family are very important to a large variety of insects. Fruits (achenes) attached to a hairy or bristly tuft are characteristic of species within this family.

HAWKWEEDS (*HIERACIUM* AND *PILOSELLA* SPECIES)

The hawkweeds have flower heads in clusters up to one inch in diameter, ray flowers only, and leaves and stems with milky sap when cut or bruised. They range in height from about six inches to two to five feet, depending on the species and site conditions. The three introduced and widely naturalized species included here are found in disturbed, sunny habitats and have leaves only in basal rosettes (versus leaves along the stem, as does the native Canada hawkweed).

ASTERS (*DOELLINGERIA, EURYBIA, OCLEMENA,* AND *SYMPHYOTRICHUM* SPECIES)

The Adirondack species of the genus *Aster* have recently been placed into the genera shown in this guide, including *Doellingeria, Eurybia, Oclemena,* and *Symphyotrichum*. All have ray and disk flowers. Besides the high value of aster flowers to many pollinators, including butterflies, bees, and wasps, seeds provide food for birds during the winter. Only the more common Adirondack asters, all native, are included here.

JOE PYE WEEDS (*EUTROCHIUM* SPECIES)

Eutrochium comes from the Greek *eu*, meaning "true" or "real," and *trocho* ("wheel-like"), referring to the leaves that are in whorls, mostly in threes to fives, depending on the species, along the stem. This genus lacks ray flowers. Joe Pye, according to folklore, was a Native American who sold these plants for medicinal purposes.

GOLDENRODS (*SOLIDAGO* SPECIES)

Goldenrods are among the most familiar of Adirondack wildflowers, often forming spectacular stands in the late summer and early fall, valued for their bright color on a dulling landscape. All but *S. bicolor*, silver rod, are yellow-flowered perennials that arise from spreading rhizomes or clumps. Goldenrods, consisting of species from the genera *Euthamia* and *Solidago*, are one of the largest groups of wildflowers in number of species. Because of their diversity and inherent variability, they often are a challenge to identify. Leaves are essential for determining most of the species. Key leaf characteristics include the place where leaves are borne (at the base—basal, or on the stem—cauline), hairiness (pubescence), and the leaf margin (toothed or smooth).

What appear to be the goldenrod blooms are actually clusters of very small flowers aggregated into a capitulum, or head. The ways in which these heads are borne—flat-topped, rod-shaped, or small clusters along the stem—are helpful criteria for identification. Fruits are superficially similar—a single-seeded dry fruit with hairlike bristles that aid in dispersal. Standing in a large patch of flowering goldenrods with numerous pollinators makes it clear that they are insect pollinated and therefore not to be incriminated as a source of hay fever, which is usually caused by pollen that is wind-borne.

Orchid Family (Orchidaceae)

When one sees the word "Adirondacks," what word next typically comes to mind? For most, it certainly is not "orchids." And yet the Adirondacks are an outstanding region in which to see about three dozen of the approximately sixty species of native orchids known to occur in New York.

If there are any charismatic plants that rival the charm and allure of the animal species that capture the public's attention, those plants are our native orchids. The orchid family is one of the largest in the world, with over twenty-five thousand known species. Worldwide, most species are tropical epiphytes, that is, they do not live in the soil but are perched in trees. All of the orchids in New York and elsewhere in temperate and northern ecological communities are terrestrial—germinating and growing in soil.

Why are there so many orchid species in the Adirondacks? Orchids occur in just about any kind of habitat, from the sunniest to the darkest, the driest to wettest, the most acidic to most alkaline, or the most to least disturbed. An area the size of the Adirondacks, with its many habitats, unique land use history, and current preservation status, offers many acres of all these conditions. Some of the best habitats for orchids are wetlands—especially acidic and alkaline peatlands (namely, bogs and fens, including northern white-cedar swamps)—and the Adirondacks have extensive areas of these wetlands in which ten or more orchid species can occur together. And given the wide range of climatic conditions encountered as one travels from the southern to northern, and western to eastern Adirondacks, as well as changes in elevational range over these distances, many of the orchid species have their northern or southern range limits in this region.

ORCHID BIOLOGY AND POLLINATION

Few families of plants have been more extensively studied in terms of insect pollination than orchids. No orchids are known to be strictly pollinated by wind or water—a fact that is reflected in the spectacular beauty of the flowers, which are structured to attract insects. We are drawn to them because of their intrinsic interest, extraordinary diversity, and exceptional beauty. But there is much more to orchids than meets the eye. They have a remarkable way of sustaining themselves through their intimate relationship with fungi. And like other underground agents, they are also masters of deception.

All orchids live in association with fungi, often species of mushrooms, in a relationship termed "mycorrhizal." This is why the roots of so many orchids have thickened tips, owing to the sheath of fungi that surrounds and/or penetrates them. Unlike most mycorrhizal relationships, however, the fungi apparently receive very little in return from the orchids, which can actually consume them. This helps to explain why some orchids can be actively growing but do not regularly emerge above ground for a period of several years. A number of orchids have become totally dependent on the fungi and produce little, if any, of their own food. The best example in the Adirondacks is the genus *Corallorhiza*. Its species have reduced chlorophyll and are not green except for *C. trifida*, which has limited photosynthesis. Their unpredictable flowering or even emergence has recently been linked with the availability of suitable mycorrhizal fungi in the vicinity of orchid roots.

Orchids are also dependent on insects. The vast majority of orchid flowers must have insect visitors in order to transfer pollen, although a few, especially species with tiny flowers or those growing in the Arctic, can pollinate themselves if no insects are available. The widespread yellow lady's slipper was the subject of a pollination study in 1862 by Charles Darwin, who noted the deceptive nature of the flower. The pollinator, usually a bee, enters the main opening in the "slipper" but becomes trapped in the sac. Its only exit is via a different route at the back of the flower, one that requires it to pass between the reproductive structures, where it picks up one of two pollen packets (pollinia).

Orchids are so experienced with insects that they have evolved ways of exploiting the pollinators without having to produce energetically expensive nectar as a reward. In fact, the flowers of many, if not most, of the orchids in our region are deceptive. This means that insect visitors are not rewarded with nectar but are tricked into thinking that nectar is present to induce a visit. Examples include the grass pink. Bees, the most frequent pollinator, are drawn to the upper lip of the flower, where there are hairs resembling stamens that the insect associates with a nectar source. However, the lip is hinged, and the weight of the misguided bee bends the lip and brings the animal in touch with the reproductive structures on the lower lip, where either the flower's pollen packet sticks to the insect's back and is then carried away, or a pollinium that the bee is carrying from a previous flower is deposited. Other genera and species of Adirondack orchids have similarly elegant mechanisms of deceptive pollination.

Orchids spare no ingenuity in exploiting insects—even mosquitoes. The pollination of several species of *Platanthera* is effected by these abundant Adirondack pests. For example, the blunt-leaf orchid is pollinated by the snow pool mosquito, so-called because its larvae favor snow pools at higher elevations. In this case, as in other mosquito pollination, the pollinium attaches to the insect's eye.

Fruits are seldom seen on orchids with deceptive pollination because they are obligate outcrossers (meaning that self-pollination does not result in seed development), and the resulting paucity of seed poses serious problems for orchid conservation. Of course, not all orchids are deceptive. Some reward pollinators with nectar. Among these are the ladies' tresses.

ORCHID CONSERVATION

Orchid conservation is of concern just about everywhere these species occur, whether temperate or tropical, terrestrial or epiphytic. Habitat alteration and outright destruction, along with human collecting have been the biggest threat for decades. Astoundingly, it is not illegal to pick or dig up native orchids in some states. More recently, browsing by white-tailed deer is wiping out populations of some orchid species, especially those that have leaves above ground level and are in places that are especially favorable to deer—for example, the showy lady's slipper (*Cypripedium reginae*) in cedar swamps. Although all native orchids in New York are protected by state law from picking and digging, there is no protection from deer browsing or the indirect effects of beaver-dam flooding of the forested conifer swamps in which many orchid species can occur.

Orchid populations are exceptional candidates for citizen science projects that follow population variation over time relative to local and regional environmental changes. Populations of the more southern and northern Adirondack species would be especially interesting to track relative to trends in precipitation and temperature. Populations that are at least half protected from browsing by installing deer exclosures can readily show the effects of browsing, forest canopy development, forest canopy gaps, or other environmental fluctuations like changes in hydrology related to beaver flooding or dam removal.

ORCHIDS IN THE GARDEN

Growing most of the native plants in this guide is covered elsewhere (see Leopold 2005). Terrestrial orchids are generally a bit more challenging, and their cultivation should not even be attempted if a legal source of plants is not available. But because there are some good options for legally purchasing many of these species, and given the books on the subject (Mathis 2005, Tullock 2005), one can seriously pursue the cultivation of native orchid species. The key is to provide not only the environmental conditions under which these species thrive in natural populations but to ensure that these conditions also support the fungal species on which that orchid species depends.

The most critical garden factors are light, soil, and moisture. Every orchid species has an ideal light regime that will maximize its flowering and overall longevity. Soils are relatively easy to amend with sand, peat moss, humus, sulfur, limestone, and so forth. Related to the substrate characteristics are the moisture regimes, that is, whether the substrate drains freely, remains saturated for extended periods of time, becomes excessively dry, and so on. The natural habitats described for these species in this book should be good general guides for what each species generally requires. References included here should be consulted for more specific information.

Native Orchid Species of the Adirondacks (covered in this guide)

Arethusa bulbosa	dragon's mouth
Calopogon tuberosus	grass pink
Corallorhiza maculata	spotted coralroot
Corallorhiza trifida	early coralroot
Cypripedium acaule	pink lady's slipper
Cypripedium arietinum	ram's head lady's slipper
Cypripedium parviflorum	yellow lady's slipper
Cypripedium reginae	showy lady's slipper

Galearis spectabilis	showy orchid
Goodyera pubescens	downy rattlesnake plantain
Goodyera repens	dwarf rattlesnake plantain
Goodyera tesselata	checkered rattlesnake plantain
Liparis loeselii	fen twayblade
Malaxis monophyllos var. *brachypoda*	white adder's mouth
Malaxis unifolia	green adder's mouth
Neottia auriculata	auricled twayblade
Neottia convallarioides	broad-lipped twayblade
Neottia cordata	heart-leaved twayblade
Platanthera aquilonis	northern green bog orchid
Platanthera blephariglottis	white fringed orchid
Platanthera clavellata	little club-spur orchid
Platanthera dilatata	tall white bog orchid
Platanthera flava var. *herbiola*	northern tubercled orchid
Platanthera huronensis	Lake Huron bog orchid
Platanthera grandiflora	large purple fringed orchid
Platanthera hookeri	Hooker's orchid
Platanthera lacera	ragged fringed orchid
Platanthera macrophylla	large-spurred round-leaved orchid
Platanthera obtusata	blunt-leaved orchid
Platanthera orbiculata	small-spurred round-leaved orchid

Platanthera psycodes	lesser purple fringed orchid
Pogonia ophioglossoides	rose pogonia
Spiranthes cernua	nodding ladies' tresses
Spiranthes lacera	slender ladies' tresses
Spiranthes lucida	shining ladies' tresses
Spiranthes romanzoffiana	hooded ladies' tresses

Other Orchid Species Known in or near the Adirondacks (not included in this guide, except in this listing)

Amerorchis rotundifolia*	round-leaved orchid
Aplectrum hyemale	puttyroot
Calypso bulbosa var. *americana**	fairyslipper
Coeloglossum viride	long-bracted orchid
Corallorhiza odontorhiza var. *odontorhiza*	autumn coralroot
Corallorhiza striata var. *striata*	striped coralroot
Isotria verticillata	large whorled pogonia
Neottia bifolia	southern twayblade
Platanthera ciliaris	orange fringed orchid
Spiranthes casei var. *casei*	Case's ladies' tresses
Spiranthes ochroleuca	yellow ladies' tresses
Triphora trianthophoros ssp. *trianthophoros*	three-birds

*no extant populations are known in New York

Trilliums *(*Trillium *species)*

Recently moved by taxonomic specialists from the lily family (Liliaceae) to the bunchflower family (Melanthiaceae), *Trillium* species are among the showiest and most beloved wildflowers of the eastern United States. Of the over three dozen *Trillium* species in North America, four can be found in the Adirondacks. *Trillium* species are typically identified first by whether their flowers have a stalk, or pedicel. All of the Adirondack *Trillium* species have stalked flowers and are between eight and sixteen inches tall.

Although none are truly rare in New York, they are all state-protected because they are exploitably vulnerable to picking and digging. They are one of the white-tailed deer's very favorite browse and are among the first wildflower species to disappear under excessive deer pressure. Most species require about six years from the time of seed germination to flowering, so they are slow to recover, if ever, after being decimated by picking or browsing.

The name *Trillium* is derived from *tri*, referring to leaves, petals, and sepals in threes, and *lilium*, for lily. All have three-part (trifoliate) leaves (technically bracts—the reduced true leaves are at the base of the plant) that persist through the entire growing season. The fruits of the four Adirondack *Trillium* species are fleshy and red to purple at maturity. *Trillium* seeds, like those of other spring wildflowers such as *Asarum canadense* and *Sanguinaria canadensis*, are attached to fleshy, fatty structures called elaiosomes, which attract ants for dispersal. Ants are apparently so excited about eating the elaiosomes that they will cut holes into the fruit before the berry ripens. The ants carry the seeds back to their nests, eat the elaiosomes, and leave the unharmed seeds in their tunnels, where they later germinate.

When legally available, some species like *T. erectum* and *T. grandiflorum* are very easy to grow in the garden if the soil is moist but well drained, fertile, and in a semi-shaded location. Others, like *T. cernuum*, which thrive on shaded hummocks in conifer swamps, are more specialized in their growing conditions.

Although they have been used historically for medicinal purposes, including as a nerve tonic, they are not considered to be edible and safe for human consumption.

Violets *(*Viola *species)*

Violets (Violaceae) are among the best known of all our wildflowers. In fact, more states claim violets as their state flower than any other native plant. Recognizing a plant as a violet is easy. However, distinguishing among the dozen or so Adirondack species can be challenging. Leaves are quite variable, and those that are formed first can be significantly different in shape and lobing from those produced later. Most violets have leaves that are generally heart-shaped.

For identification purposes, violets can be placed into several categories. First is the growth form. If there is no evident stem and the leaves and flowers appear to arise directly from the root, the violet is stemless, or acaulescent (without a stem). Contrariwise, if an obvious stem is present, the violet is caulescent (with a stem).

Violets produce blue, yellow, or white flowers in the spring or early summer. Occasionally, a white-flowered individual will be found in a population among its blue kin. Flowers of violets are borne nodding singly on the flowering stem and have wings that are their signature feature. One of the unusual aspects of violet biology is the production of a second type of flower, often overlooked. These are the cleistogamous (hidden reproduction) flowers that are usually borne near the base of the plant and in some species are actually subterranean. These flowers never open and are self-fertilizing. Open, aerial flowers are chasmogamous (open reproduction) and frequented by bees. The early yellow violet, *V. rotundifolia*, is stemless during spring flowering, but it also produces a prostrate stem with one to three cleistogamous capsules in the summer. Both types of flowers produce seeds, but cleistogamous flowers are more fecund. By having a mix of cross-pollinated (chasmogamous) and self-pollinated (cleistogamous) flowers, violets are able to maintain genetic diversity. Many violets also reproduce by runners (stolons).

Species Accounts

Plant descriptions are organized by predominant flower color. Within color groups, species are arranged alphabetically within families, which are also listed alphabetically.

Yellow to Orange Flowers

Pastinaca sativa L., wild parsnip (Apiaceae)

A rogue relative of the root vegetable parsnip, wild parsnip has morphed into one of the most serious dermatitis-causing plants in the nation, and it is found throughout much of northern New York. It is especially abundant in the Adirondacks along roads. Its origin is unknown; it could be the result of natural selection in the cultivated parsnip, or it could be a more recent introduction from Europe.

Plants are up to five feet tall with grooved stems and compound leaves. The entire plant when crushed (but don't crush it without gloves) smells like celery or carrot. Like its other relatives in the parsley family, it has flat-topped inflorescences (flower clusters). Wild parsnip has masses of small yellow flowers in midsummer. Fruits are flattened and resemble fennel seed.

All parts of the plant contain compounds that cause phytophotodermatitis induced by contact with the plant. These irritants are absorbed into the skin and when activated by light, initiate a chemical transformation causing swelling and blisters. This reaction is not an immune response but rather a contact poison, so it operates in a manner different from that of poison ivy.

Wild parsnip is difficult to control because of its large, persistent root, which requires digging for effective removal.

Zizia aurea (L.) Koch,
common golden Alexanders (Apiaceae)

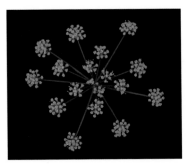

Golden Alexanders occurs in rich, mesic forests (those with a moderate amount of moisture), in floodplains, and moist fields. Individual flowers are bright yellow, one-eighth of an inch in diameter, and arranged in a flat cluster (an umbel) that resembles spokes of an umbrella and is about one to two inches in diameter. Leaves are doubly compound and consist of nine to twenty-seven toothed leaflets. Plants often grow in patches and can reach two feet in height. At first glance, this species might be confused with the wild parsnip, *Pastinaca sativa*, but the latter species is taller, flowers later, and has more leaflets per compound leaf.

A homeopathic medicine is derived from the root. Golden Alexanders is a host plant to black swallowtail butterfly larvae. The genus is named after J. B. Ziz (1779–1829), a German botanist.

Asclepias tuberosa L., butterfly weed (Apocynaceae)

Unlike its milkweed relatives, butterfly weed has alternate leaves and lacks milky latex (juice). It is easily recognized by its bright orange flowers that appear throughout most of the summer. Butterfly weed is also different from most other milkweeds in that it can branch from the base and produce a solid clump of many flowering stems. Its ease of cultivation and attractive, long-blooming flowers make it a popular garden subject, especially for anyone interested in monarch butterflies whose larvae feed on the plant.

The alternate leaves are usually hairy. Fruits are erect, narrow, and often densely hairy. Like other members of the genus, the flattened seeds are tethered to long, silky hairs that enable the seeds to be lofted in the air.

Hemerocallis fulva L., orange day lily (Asphodelaceae)

The orange day lily often forms dense patches along sunny roadsides, borders of fields, and at old homesites. Even though its flowers do not produce fertile seeds (plants reproduce by vegetative means) it is widely naturalized from Europe via Asia and has long been in cultivation.

It has six flower parts (three petals and three petal-like sepals in this species). The plant blooms in summer, and the individual flowers, three to fifteen per stalk, last only one day. These

rather large (up to five inches in diameter), funnel-shaped orange flowers with a deep yellow throat are held upright above
straplike leaves that are about two feet long. *Hemeros*, is the
Greek word for "day," and *kallos* means "beauty," referring to
the single-day longevity of each flower once it opens.

Its edible buds, flowers, and tubers can be boiled, fried,
baked, or pickled. The petals and sepals taste like lettuce.

Bidens cernua L., nodding beggar ticks (Asteraceae)

B. connata Muhl. ex Willd., purple-stemmed beggar ticks (Asteraceae)

Bidens frondosa L., devil's beggar ticks (Asteraceae)

About a half-dozen native beggar ticks occur in the Adirondacks. These annuals are found in a variety of open wetlands and disturbed wet sites, with saturated soils under sunny conditions being the common habitat characteristics. All bloom in late summer and typically reach about three feet in height.

Flower heads of *B. cernua* are up to two inches in diameter and consist of about eight bright yellow rays around the central brownish-yellow disk flowers. Heads tend to droop, hence the common name. Leaves are entire (that is, not lobed), lance-shaped, about five inches long, and opposite one another on the stem.

Flowers of *B. frondosa* lack ray flowers and thus appear to have no "petals." They are about one-third of an inch wide and consist of yellow disk flowers and five to nine leafy bracts. Leaves are pinnately compound, consisting of three to five leaflets arranged in a row. *Bidens connata* is very similar, except the leaves are not compound and stems are often purplish.

Bidens means "two-toothed," referring to the two-pointed barbs characteristic of the dry fruit of some species of this genus, including *B. frondosa*. Other species, like *Bidens cernua*, have two to mostly four barbs per fruit. These barbs readily attach to animal fur and clothing to facilitate seed dispersal.

Crepis capillaris (L.) Wallr.,
smooth hawk's beard (Asteraceae)

Similar in appearance and habitats to the hawkweeds, the smooth hawk's beard is found in pastures, along scraped roadsides, and in other disturbed sites in full sun. This annual or biennial was introduced from Europe and is becoming more widely naturalized. Leaves are narrow and usually toothed or lobed. Flower heads are less than an inch in diameter, consisting of up to sixty ray but no disk flowers, and are present from July through September. Its young leaves and shoots are edible.

Euthamia graminifolia (L.) Nutt., flat-topped goldenrod (Asteraceae)

As the scientific name implies, the leaves are grasslike in this goldenrod. Like other members in the genus *Euthamia*, flat-topped goldenrod has small heads with fifteen to twenty flowers. In addition to their shape, the leaves are distinctive in having abundant clear spots from oil deposits, which are clearly evident when holding a leaf up to the light.

Flat-topped goldenrod is common throughout the Adirondacks in moist fields, wet meadows, and roadsides. It flowers in late summer.

Helenium autumnale L., sneezeweed (Asteraceae)

Sneezeweed occurs in open wetlands, including marshes, wet meadows, and streambanks. Flower heads are about two inches wide with thirteen to twenty-one three-lobed, skirtlike rays that turn back, and a very prominent yellow, nearly globular disk. Flowering occurs in late summer into autumn (*autumnale*), with up to one hundred flower heads per branched cluster. Leaves are alternate, narrow, up to six inches long, winged and dotted; stems are winged. Plants can reach five feet in height.

The common name for this plant can possibly be attributed to its being erroneously blamed for causing hay fever, which is actually caused by ragweed species (*Ambrosia trifida* L. and *A. artemisiifolia* L.) that bloom at the same time but are rather inconspicuous. Sneezeweed pollen is dispersed by insects, not wind. Another source indicates that the disk flowers and leaves were dried, crushed, and used as snuff. Plant parts are poisonous if eaten in large quantities.

Hieracium kalmii L., Canada hawkweed (Asteraceae)

The Canada hawkweed is the only native hawkweed listed here. It occurs along roadsides, in meadows, and in dry upland forests. Flowers are yellow, opening in summer until fall. Leaves are toothed and attached directly at their base along the stem. Plant parts are white-hairy.

This plant was named by Linnaeus for his student, Pehr Kalm, who botanized in the Adirondacks. The species name for Canada hawkweed is listed as *Hieracium canadense* or *H. umbellatum* in some guides.

Inula helenium L., elecampane (Asteraceae)

Elecampane is a robust species from Eurasia and naturalized in the eastern and western United States, occurring in moist to wet meadows, along roadsides, and in other sunny, disturbed sites. It is considered invasive in some states. Flower heads reach four inches in diameter, open in summer, and consist of fifty to one hundred yellow narrow ray and disk flowers. Leaves are alternate, large, up to twenty inches long, and densely hairy on the underside. Plants can reach seven feet in height.

This species has been known for millennia for its medicinal and food uses, and has been long valued to treat chronic bronchitis and bronchial asthma. Recent research suggests efficacy of elecampane as an effective antibiotic. The root is also used as a flavoring in absinthe. The most prominent constituent is inulin, an indigestible carbohydrate known as a fructan. Inulin is produced commercially from chicory (*Cichorium intybus*), which is related to elecampane.

Lactuca canadensis L., tall lettuce (Asteraceae)

Lactuca biennis (Moench) Fernald, tall blue lettuce

Tall lettuce is a native annual/biennial of moist fields, meadows, and open woods. Flower heads open from midsummer until fall, are three-eighths of an inch wide, consist of thirteen to twenty-two pale yellow to purple ray flowers, and lack disk flowers. Flower heads occur in large, open clusters that resemble a candelabra. Leaves are prickly and highly variable in shape. Plants can reach eight feet in height.

The tall blue lettuce, *Lactuca biennis* (Moench) Fern. also occurs in the Adirondacks in similar habitats. Flower heads are similar to tall lettuce, but rays are bluish to white. Leaves have winged stalks.

Stems and leaves of all *Lactuca* species exude milky sap when torn (*lac*, "milk"). Young leaves of both species are edible in salads or cooked, although they are somewhat bitter.

Packera aurea (L.) Á. & D. Löve, golden ragwort (Asteraceae)

Golden ragwort occurs in moist to wet sites, in open to closed canopy habitats, and often in dense patches. It is one of the earliest blooming members of the composite family, flowering from midspring until early summer. Its flowering heads of ray and disk flowers are yellow to golden, in heads about three-quarters of an inch in diameter. Basal leaves are oval, with a heart-shaped base, and much larger when the habitat is shady. Leaves along the stem are pinnately lobed (with the lobes arranged opposite each other on either side of a central axis). Plants are relatively short in full sun, about one foot tall, but can reach three feet, especially in moist, shaded habitats.

In many guides, this species is listed as *Senecio aureus*. The species name comes from the word *aur*, Latin for "gold." The more recent genus name, *Packera*, is after John Packer, a twentieth-century biologist. Some guides list the common name as "groundsel."

Pilosella aurantiaca (L.) F. W. Schultz & Sch. Bip., orange hawkweed (Asteraceae)

The orange hawkweed was introduced from Europe and is naturalized in the United States, occurring in dry fields, pastures, roadsides, and other disturbed, open sites. Flower heads of red to orange ray flowers up to one inch in diameter open from summer until early fall. Its leaves are basal, and its stems are very hairy. Plants reach two feet in height. Its roots form mats.

The genus of this species is listed in many guides as *Hieracium aurantiacum*. Another common name for this species, Indian paintbrush, is best reserved for *Castilleja coccinea* (L.) Spreng., native to New York but absent from the Adirondacks.

Pilosella caespitosa (Dumort.) P. D. Sell. & C. West, yellow hawkweed (Asteraceae)

Yellow hawkweed is similar to the next species, except yellow hawkweed stems are very hairy and its roots form mats via long runners. It is widely naturalized along roadsides, in old fields, meadows, and pastures. Flowers are bright yellow, in one-half-inch-diameter heads in short-stalked, compact terminal clusters on a tall stalk, from midspring until late summer. Flower stalks hold from five to thirty heads. Leaves are in basal rosette flat on the ground and similar to the next species but green. Plants can reach two feet in height.

The genus for this species is listed in many guides as *Hieracium*.

Pilosella piloselloides (Vill.) Soják, glaucous-leaved hawkweed (Asteraceae)

This species is also known as glaucous king devil. It has a smooth (hairless) stem and its roots do not form mats via runners. It is naturalized along roadsides, in old fields, meadows, and pastures. Flowers are yellow and are in long-stalked heads in a more open cluster than the previous species. Leaves are grayish-green in color (glaucous) and basal. Plants can reach over three feet in height.

The genus for this species is listed in many guides as *Hieracium*.

Rudbeckia hirta L., black-eyed Susan (Asteraceae)

One of the most familiar of all roadside and old field wildflowers, black-eyed Susan is not native to New York but has become widely naturalized from midwestern prairies in sunny, disturbed areas. Blooming from summer into fall, the flower heads are about three inches in diameter and consist of eight to twenty-one bright yellow rays around a chocolate-colored, domed disk. Leaves and stems are rough-hairy. Plants are biennial or short-lived perennials and can reach about three feet tall.

Linnaeus named this genus after his botany professor, Olaf Rudbeck (1660–1740).

Rudbeckia laciniata L.,
green-headed coneflower (Asteraceae)

The green-headed coneflower is a tall (to ten feet) native plant of open, moist meadows and wetlands, including floodplains and streambanks. Flower heads are up to four inches in diameter and consist of six to sixteen yellow, drooping, straplike ray flowers around a yellow-green to gray knoblike disk. Flowering occurs from midsummer until fall. The species name *lacinia-ta* ("slashed" or "lacerated") refers to the deeply divided, large lower leaves, which are up to twelve inches long and wide, and have three to seven lobes from the base up to the middle of the stem. Plants often occur in patches.

Solidago caesia L.,
blue-stemmed goldenrod (Asteraceae)

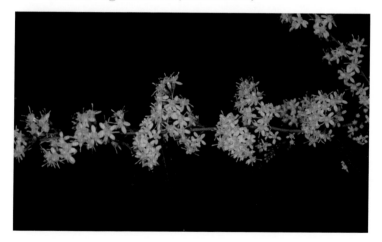

Caesia means "blue gray," an accurate description of the stem, at least when the plant is young. Plants usually have a single stem. Blue-stemmed goldenrod has a round-in-cross-section (terete) stem and cauline (borne on the stem) leaves. Likewise, the clusters of flowers arise at the junction (axil) of the stem and the upper leaves. This goldenrod is able to survive in deep shade and grows in moderately moist to moderately dry upland forests.

Solidago canadensis   *Solidago gigantea* Aiton,
L., Canada goldenrod   swamp goldenrod
(Asteraceae)

Among the tallest of our goldenrods, Canada goldenrod is frequent in the Adirondacks in fields and roadsides, where it spreads aggressively by rhizomes. It is large, with many crowded leaves, and it flowers in the late summer and fall from clustered green stems that are downy above the middle. The masses of flowers are typically borne on drooping upper branches. They are all carried on only one side of the flowering branches (secund).

This species is very similar to *S. altissima*, tall goldenrod, and some botanists lump the two species together. It is also very similar to the swamp or smooth goldenrod, *Solidago gigantea* Aiton, which has a smooth, whitish stem and is found on wetter, often shaded sites. Both are also widespread in the

Adirondacks. All three can exceed six feet in height, with *S. gigantea* reaching eight feet.

Solidago flexicaulis L., zigzag goldenrod (Asteraceae)

So-called because of the arrangement of the leaves and flower clusters on the somewhat zigzagging stem, this goldenrod prefers upland forests in deep shade and is often found along with *S. caesia*. Flower clusters are in the axils of the leaves as well as in a short, terminal flowering stem. Leaves are distinct, with large teeth and a tapered tip, not conspicuously hairy. Zigzag goldenrod favors base-rich (alkaline) soil substrates.

Solidago hispida Muhl. ex Willd., hairy goldenrod (Asteraceae)

Hairy goldenrod occurs in dry woods and meadows, and is often associated with calcium-influenced soils and bedrock. As the common name suggests, all plant parts are densely hairy (*hispida*, "with stiff hairs"). Its spikelike flower clusters are present from summer until fall. Plants can reach about three feet in height.

Solidago juncea
Aiton, early goldenrod
(Asteraceae)

Solidago rugosa Mill.,
wrinkle-leaved goldenrod
(Asteraceae)

These medium-sized goldenrods with pyramidal inflorescences are common old-field, pasture, and open woodland species. Leaves and stems of *S. juncea* are smooth versus hairy or rough for *S. rugosa*, and the leaf margins of *S. rugosa* are coarsely toothed. *Solidago juncea* flowers in midsummer, whereas *S. rugosa* flowers later in the summer into fall. Early goldenrod thrives on drier sites, while wrinkle-leaved goldenrod favors wetter ones.

Solidago leiocarpa DC.,
Cutler's alpine goldenrod
(S2) (Asteraceae)

Solidago macrophylla
Banks ex Pursh, large-
leaved goldenrod
(Asteraceae)

These two goldenrods are the most "mountainous" of the Adirondack goldenrod species. The large-leaved goldenrod, with its broad leaves and half-inch diameter flowers, occurs in canopy openings, especially of subalpine forests. Cutler's alpine goldenrod is found at the highest elevations in the alpine community. Although relatively frequent in this region, it is state-protected because of its overall rarity in New York.

Solidago nemoralis Aiton, gray goldenrod (Asteraceae)

Dry, infertile sites favor this patch-forming goldenrod, especially in old fields, roadsides, and waste places—all in full sun. This site affinity is ironic, since the Latin *nemoralis* refers to woodlands. It is widespread, flowering in late summer. Stems are solitary or tufted. The short, soft hairs, especially on the larger and lower leaves, look like dust, giving them a gray cast and providing the common name. Leaves have irregular teeth, with the upper and smaller cauline leaves (borne on the stem) lacking teeth. The flowering stem is rather narrow and wandlike.

Solidago patula Muhl. ex Willd., rough-leaved goldenrod (Asteraceae)

The flowering branches of this tall goldenrod have been described as "messy" because they extend in all directions. The largest leaves are near the base, becoming smaller all the way to the top of the stem up to just below the flowering branches. Found in open fens and wet meadows, this robust plant, which grows up to seven feet tall, is covered with short, stiff hairs that feel like sandpaper. The wings on the stem are another distinctive feature.

Solidago uliginosa Nutt., bog goldenrod (Asteraceae)

The Latin name refers to marshes, the common name for bogs. But this tall (to five feet in height), single-stemmed goldenrod favors sunny fens and only rarely is found in bogs. The flowers tend to be borne on only one side of the flowering branches (secund).

Tragopogon pratensis L., meadow salsify (Asteraceae)

Tragopogon dubius

Also known as goat's beard, meadow salsify is widely naturalized from Europe, and in the Adirondacks it occurs in old fields, along roadsides, and in other sunny, disturbed sites. Flowers are about two and one-half inches in diameter, open in spring to early summer, and consist of deep yellow ray flowers; disk flowers are lacking. Flowers close by noon. One-seeded fruits are in a large ball-like cluster, resembling those of dandelions but much larger (up to three inches in diameter). Leaves are grasslike and clasp the stem. Plants reach about two feet in height. Plant parts exude milky sap when torn.

Species of this genus, including meadow salsify, are often referred to as goat's beard because of their large, fuzzy seed heads. Young roots can be eaten raw; older roots should be cooked in the same manner as parsnips. Plant parts have a high inulin content and have therefore been used for medicinal purposes. The fistulous salsify, *Tragopogon dubius* Scop., occurs in New York mostly outside of the Adirondacks. It differs from *T. pratensis* in having very long flower bracts that extend way beyond the ray flowers.

Tussilago farfara L., coltsfoot (Asteraceae)

Coltsfoot is naturalized from Europe and occurs in highly disturbed areas like roadsides, eroding streambanks, and waste places. Flowers are among the first of all species to appear each spring, emerging from bare ground without leaves and in dense patches. Flower heads are solitary, about one inch in diameter, and consist of many narrow, yellow ray flowers around the yellow disk flowers, which are borne at the top of a stout, scaly stalk. The leaves fully emerge after the seeds are dispersed and are heart-shaped and densely white-hairy beneath.

The genus name comes from the Latin, *tussis*, for "cough"; ground leaves and stems have been used for centuries, especially in Chinese herbal medicine, to treat coughs and other chest ailments. Some herbal cough syrups still use this species. The plant contains traces of an alkaloid that can be toxic to the liver in large doses; these toxins are minimized when the plant is boiled.

Impatiens capensis Meerb.,
spotted jewelweed (Balsaminaceae)

Impatiens pallida

Impatiens sounds like "impatience" and has the same root. It refers to the seed capsules, which suddenly rupture and fling seeds without warning. The specific epithet refers to the Cape of Good Hope, because early botanists who saw it in Europe thought it came from there. Jewelweed is so named because when the leaves are held under water, the wax on their surface repels the water giving the leaf a silvery sheen. It is an annual but can grow up to six feet tall in a season, largely through the rapid accumulation of water evident in the very succulent stems. The common reference to the use of the stem juice as an antidote for poison ivy has a scientific basis. The plant's leaves contain small concentrations of saponins (compounds that produce a soapy lather), which have a detergent action that could help wash the urushiol in poison ivy off the skin before it can cause irritation. Crushed whole stems are more effective than an extract from the stems.

Wetlands such as streambanks, seeps in forests, and other shaded wet areas are the habitat for jewelweed, as well as dense freshwater marshes. The bright-orange flowers are produced all summer and dangle delicately below the leaves, where they are obvious to the pollinators—bees, butterflies, and hummingbirds. Also widespread and often growing with spotted jewelweed is *Impatiens pallida* Nutt., pale jewelweed. Habitats of the two species are similar, although pale jewelweed is more common in heavily shaded conditions. Both jewelweeds share pollinators and thus have evolved a system where the pollen of one species is physiologically incompatible with the other.

Flowers of jewelweed are edible, but the stems and leaves should be avoided because of their low levels of saponins, which are toxic.

Caulophyllum thalictroides (L.) Michx., blue cohosh (Berberidaceae)

If you find a stand of blue cohosh, you are probably in one of the richest assemblages of spring ephemerals. It only thrives in fertile, mesic (moderately moist) forests and is an excellent indicator of the quality of such communities.

This hardy perennial grows up to three feet tall with a single stem. Leaves have a purplish-blue tint when unfolding in spring. At flowering there are two leaves. Each leaf is divided into three leaflets that are again divided into three segments. The overall leaf is therefore quite broad. Leaf and stem surfaces are often gray or gray green because of the wax found on some plants. Flowering begins before the leaves are fully formed. What appear to be yellow-green or yellow-purple petals are actually the sepals. Petals are very reduced and modified into nectar-secreting structures (nectaries). Like all members of the barberry family (Berberidaceae), the anthers are opened by flaps to release the pollen.

Blue cohosh has a fruit different from any other in our flora. As the fruit develops, the ovary ruptures and withers, and the two seeds develop without a fruit coat—more like a gymnosperm than a flowering plant. The colorful blue seeds, like the

rest of the plant, are toxic. This is one reason why it is not subject to severe damage by grazing deer.

The name "cohosh" is thought to be derived from an
Algonquin word for a medicinal plant or group of plants.

Barbarea vulgaris Aiton,
yellow rocket (Brassicaceae)

This plant is native to Europe,
where it is a winter annual—a
species that germinates in
mild winter weather, forming
a rosette of basal leaves from
which, in the second year,
the flowering stems arise.
Obviously, it does not germinate in the Adirondack winter. Rather, it germinates in
the early spring. Yellow rocket
is so called because its leaves
bear a resemblance to those of
the vegetable rocket (from the French *roquette*), better known
in English as arugula (*Eruca sativa* L.), and it has a somewhat
similar taste. Growing to about three feet tall and often with
multiple stems, it begins to flower in early spring and continues into midsummer with masses of mustard-yellow flowers.
These have four petals and yield a long, thin fruit with a row of
seeds on each side of a partition. Fruits are usually in an angled,
upright position and mature as the plant continues to flower.

Because of its resistance to some insects and also its ability to yield an edible crop, trial fields of yellow rocket are being planted and studied to determine its potential to limit pests and produce food.

Uvularia grandiflora Sm.,
large-flowered bellwort (Colchicaceae)

The large-flowered bellwort occurs in rich, dry to moderately moist upland hardwood forests, especially where calcium exerts an influence. Flowers are bell-shaped, one to two inches long, pale to bright-lemon yellow, six-parted, twisted and nodding, opening in midspring. The fruit is a three-lobed pyramidal capsule. Leaves entirely clasp the stem and have a wilted appearance during flowering. Plants can reach twenty inches in height.

Uvularia refers to the likeness of its hanging flowers to the uvula in one's throat. Consequently, plants were once believed to be an effective treatment for throat disorders.

Uvularia sessilifolia L., wild oats (Colchicaceae)

Wild oats, also known as sessile-leaved bellwort, flowers in the spring in dry upland hardwood forests. Compared with *U. grandiflora*, this species has smaller stature (up to about twelve inches tall); smaller and paler flowers (cream-colored), single or in pairs; flower parts not twisted; and leaves not surrounding the stem. The fruit is a strongly three-winged capsule. *Sessilifolia* is Latin for stalkless or "sessile," leaves.

Hypopitys monotropa Crantz,
yellow pine sap (Ericaceae)

There is taxonomic confusion concerning the two species of
pine sap, which in older literature are subsumed under the name
H. monotropa. They are plants lacking chlorophyll and therefore lacking typical leaves. The stem is up to fifteen inches tall
and covered by scales. The younger stem is slightly drooping,
with crowded flowers about an inch long. As the plant matures,
the flowers bend upward, producing a capsule filled with thousands of dustlike seeds. The two species are easily separated on
the basis of plant color, as the common names indicate. And the
phenology (periodic biological phenomena that are correlated

with climatic conditions) is different. Yellow pine sap appears in the spring and early summer; red pine sap, in the late summer and fall.

Pine saps, like the related Indian pipe, are mycoheterotrophs—plants that lack chlorophyll and obtain their food via fungi that are associated with trees (mycorrhizal fungi). Yellow pine sap is quite host specific, favoring fungi that live in the roots of willows or pines. In fact, *Hypopitys* means "below pines." The relationship to pines is also reflected in the common name. While little research has been done on these mycoheterotrophs, it appears that *H. monotropa* requires, for its biological survival, species of the mushroom genus *Tricholoma*, which includes the well-known matsutake mushroom.

Melilotus officinalis (L.) Lam., yellow sweet clover (Fabaceae)

Melilotus albus

This introduction from Eurasia, a biennial, is widespread throughout all of the United States and Canada. It was introduced as a forage crop and to restore fertility to worn-out soils because it adds nitrogen to the soil. Like most plants with weedy tendencies, this legume grows in a variety of habitats in the Adirondacks, including old agricultural fields, roadsides, railroad rights-of-way, and sand and gravel banks of streams.

Plants are up to seven feet tall, much branched, with alternate three-parted leaves. The yellow, fragrant flowers are produced from spring until frost. They are a favorite food plant for honeybees, and the honey from yellow sweet clover is highly valued.

The closely related white sweet clover, *Melilotus albus* Medik., flowers somewhat later than its yellow-flowered relative. The two can grow together and are easily separated on flower color alone. White sweet clover has slightly larger fruits.

Hypericum ellipticum Hook., pale St. John's wort (Hypericaceae)

Flowers are paler than those of other members of the genus, like the common St. John's wort, and are a bright yellow. Flowers are five-parted and, like its kin, the stamens are in bundles. Plants are found in open wetlands and along shores, where they grow to one and a half feet tall with opposite, narrow leaves that clasp the stem. In most plants, the leaves are bent upward. The fruit is a capsule that is bright purple before opening to release the numerous seeds.

Hypericum mutilum L.,
dwarf St. John's wort (Hypericaceae)

This is the most common of the approximately ten species of St. John's worts to be found in the Adirondacks and the most frequent in wetlands. It is a short-lived perennial or annual, and like all St. John's worts has opposite leaves and bright yellow flowers. It is abundant along sandy shores of lakes, streams, and marshes. This is one of the smallest of the *Hypericum* species, seldom more than eighteen inches tall. Fruits are an attractive purple when ripe.

No nectar is produced by the flowers, so insects, mainly bees and flies, collect the pollen that attracts them. Like other members of the genus, it contains the phytophotodermatitis agent hypericin.

Hypericum perforatum L.,
common St. John's wort (Hypericaceae)

The bright yellow flowers of this European plant are the reason it was introduced as a garden subject. It is an aggressive plant that spreads from rhizomes to produce large clumps that are difficult to control. Often abundant in dry fields, roadsides, and other open disturbed areas, it grows about two feet tall, is much branched, and has opposite leaves. The leaves have numerous clear spots (actually, oil glands) in them, so when a leaf is held up to the light, it appears perforated, hence the specific epithet *perforatum*. Flowers are five-parted, and the petals often have black dots along their margins.

There has been considerable interest in using extracts of St. John's wort for a variety of ailments, including depression. Caution, however, is recommended before ingesting any part of the plant because of the presence of chemicals that cause phytophotodermatitis. This occurs when chemicals enter the body and are then affected by sunlight, causing a change in the chemical that results in burning, swelling, or blisters on the skin. The effect of the compounds, often hypericin, is termed "hypericism."

Utricularia cornuta L.,
horned bladderwort (Lentibulariaceae)

Look for this common inhabitant of shallow water at lakeshore margins, ponds, sedge marshes, and in ditches in midsummer,

when it is in full flower. All species of bladderworts lack true leaves. The large (up to three inches), brilliant, yellow flowers of horned bladderwort resemble an orchid. The long spur on the flower accounts for the common name of horned bladderwort and would appear to be an adaptation for specialized insect pollination. However, horned bladderwort is self-fertile and not dependent on insects for seed production.

There are five species of bladderworts in the Adirondacks, and all have specialized bladders for trapping and digesting small invertebrates. Bladders are submersed in the shallow water or wet soil. Their size varies from species to species—horned bladderwort has one of the largest, about one-quarter of an inch across. Each bladder, which the plant actively keeps empty of water, is fitted with a trapdoor that is surrounded by trigger hairs. When these are tripped by curious animals, perhaps mistaking the hairs for food, the trapdoor opens, and the animal is sucked inside by the resulting negative pressure. The door immediately closes behind it, imprisoning the unwary, doomed animal as a meal for the plant. The entire operation is accomplished in about 0.5 milliseconds. Lining the inner wall of the bladder are glands that digest the animals, providing

much-needed nitrogen, because bladderworts, like other carnivorous plants, live in nitrogen-poor habitats.

Clintonia borealis (Ait.) Raf., blue bead lily (Liliaceae)

The blue bead lily, also known
as corn lily, is one of the most
abundant wildflowers of the
Adirondacks, occurring in
rich, moist, forested uplands
that are often densely shaded and over a very wide elevational range. It is also common
on elevated microtopography
(hummocks or mounds) in conifer swamps.

Its yellow to greenish flowers occur in nodding clusters of
three to six in mid- to late spring. The dark green, glossy, oblong basal leaves, in groups of two to five, are sometimes mistaken for the similarly shaped leaves of the pink lady's slipper
(*Cypripedium acaule*), but the leaves of this latter species are
not glossy and are distinctly ribbed.

The genus is named after DeWitt Clinton (1769–1828),
naturalist and New York governor; *borealis* means "northern."
The common name refers to the dark blue berry fruit that is
held upright at maturity. Chipmunks and birds eat the berries.
Native Americans made a poultice from the leaves, applying it
to wounds.

Erythronium americanum Ker Gawl,
yellow trout lily (Liliaceae)

The yellow trout lily often forms carpets of beautifully mottled, lance-shaped leaves and yellow flowers in early spring in mesic (moderately moist) upland forests. A true spring ephemeral, its abundance can only be appreciated if one is out in the woods before the tree canopy closes in midspring.

Its common name refers to brownish splotches on the leaves that resemble the mottling of brook trout. Others claim that the plant is so named because it typically flowers about the time that trout season opens, or at least when trout are more likely to be caught. Flowering individuals usually have two basal leaves, with a single flower of six parts held above them, up to about six inches in height. The fruit is a capsule. Ants are attracted to the seeds and disperse them. Like many species in this family, once seeds germinate, it typically takes about six to seven years for the plant to flower. Sizable and dense patches of yellow trout lily result from spreading by short, underground stems.

Lilium canadense L., Canada lily (Liliaceae)

The Canada lily occurs in moist meadows, streamsides, edges of moist woods, and sunny openings in swamps. Under the best conditions, including high levels of sunlight and lack of deer herbivory, it can occur in large numbers.

Stems with whorls of four to twelve leaves can reach about six feet in height and typically hold one to five nodding flowers that are orange, reddish-orange, or yellow in summer. Some individuals have been found with over twelve or more flowers. Flowers are only slightly recurved and are pollinated by ruby-throated hummingbirds, honeybees, butterflies, and leaf-cutting bees. The fruit is a three-valved capsule that is held upright at maturity.

Although not rare in New York, it is state-protected because of its beauty, so although its flower buds and roots (bulbs) are edible and were once collected for food, such harvesting is now illegal.

Medeola virginiana L.,
Indian cucumber root (Liliaceae)

Indian cucumber root has short, white rhizomes in midsummer that are crunchy, with a hint of cucumber taste. A well-known wild edible, it takes a large number of plants to make a very small meal, with the resultant destruction of a population of this widespread, summer-flowering perennial. Deer, on the other hand, devour the entire plant, and in areas with large deer populations this delicate, symmetrical plant is decimated. All parts of the plant are in threes or multiples of three.

On a stem about four or five inches tall, Indian cucumber bears a single whorl of usually six leaves. The flowering stem arises two inches above the whorl with another whorl of three leaves with the drooping flowers, often three, in the center. Flowers are also in parts of three, with a prominent three-parted style, six stamens, and three greenish-yellow petals and sepals. Total plant height of flowering individuals can reach two to three feet. The fruit is a purple berry.

Nuphar variegata Durand,
common yellow pond lily (Nymphaeaceae)

This is the most common of several species of yellow water lily in Adirondack lakes and streams. It is often found growing with the fragrant white water lily (*Nymphaea odorata*), though characteristically in deeper water. Leaves and flowers arise from massive rhizomes in water seldom deeper than six feet. The leaves, in contrast to those of the white water lily, lack auricles (structures that resemble earlobes). Yellow flowers about one and a half inches wide and often tinged with red, are borne singly on a stiff stalk, resilient in water currents. Flowering takes place over several days. On the first day, a diversity of insects, chiefly beetles, visit the flower. At night, the petals recurve to trap the insects, releasing them when they open in the morning.

Fruits develop above water and contain sacs of hard, brown seeds that are edible and can be popped and eaten like popcorn.

Oenothera biennis L.,
common evening primrose (Onagraceae)

A coarse biennial or perennial up to six feet tall with alternate, simple leaves often covered with short hairs. Flowers are large, bright yellow, and visited by a diversity of insects, especially sphinx moths but also bees and hummingbirds. Flowers are only fully open at night, though on cloudy days they may remain open. Otherwise, the corolla withers after one day. It flowers in mid- to late summer and favors well-drained sites such as roadsides, as well as gravel and sand bars of streams.

Though native, it often behaves as a weed, in that it is found in disturbed areas, especially waste places.

The fruit is a narrow capsule packed with tiny seeds. The oil from the seeds is a valued component of some medicines and cosmetics. More commonly used is the sizable taproot, which forms in the first year of growth, along with a rosette of leaves. The next season the stem develops, usually unbranched but sometimes with several side branches. At this stage, in mid-summer, the roots can be collected and either eaten raw or cooked. They have a distinct odor (repugnant to some people) and a taste reminiscent of the vegetable salsify, species of both *Tragopogon* and *Scorzonera*.

Cypripedium parviflorum Salisbury, yellow lady's slipper [EV] (Orchidaceae)

The yellow lady's slipper is highly variable but unmistakable in appearance. Its nomenclature is quite confusing. Commonly

listed as *C. calceolus* for many decades, three varieties of
C. parviflorum are known in New York, all likely in the
Adirondacks. The variety *makasin* (northern small yellow lady's slipper) occurs in calcareous rich fens and cedar
swamps; *parviflorum* (southern small yellow lady's slipper) in
rich lowland forests; and *pubescens* (large yellow lady's slipper) in rich, moderately moist, upland forests. The common
characteristics among the habitats of these varieties are
a substrate that is typically calcium-rich and moisture
sufficient to not limit growth. The differences among varieties include flower size, coloration, and fragrance, as well
as other more subtle characteristics. They hybridize freely
when they co-occur. The yellow lady's slipper blooms in late
May to early June. Plants often grow as multiple stems and
are scattered in upland stands. In cedar swamps, numbers
of clumps of many stems may be encountered, resulting in a
truly spectacular show.

Dermatitis, caused by the hairiness (pubescence) of the
stems and leaves, is known to result from contact with this
species, as well as with the showy lady's slipper (*C. reginae*);
this reaction is uncommon in the other species of lady's slipper.

Of all the lady's slipper species, this species is the easiest to
grow in gardens, but only nursery-propagated plants should be
purchased, and wild plants should never be dug on private lands.
Because it is relatively easy to produce many plants through tissue culture propagation, finding legal sources of this species
is not difficult.

Pedicularis canadensis L.,
wood betony (Orobanchaceae)

Patches of this perennial occur in open woodlands. While the origin of the term "betony" is obscure, the genus name *Pedicularis* is from the Latin *pediculus* ("louse"), chosen because of the plant's reputation of infesting cattle with lice when the plant is grazed.

Wood betony is a parasite, though you would not know this at first glance. The plant is green and leafy. Yet it connects to the roots of neighboring plants to withdraw water and nutrients. This is evident in some large patches of the parasite, where surrounding grasses and herbaceous vegetation looked stunted, an effect of the parasitism. Wood betony also attaches to tree roots. Like most of these green parasites it is omnivorous—many different herbaceous and woody plants can be parasitized.

Leaves are opposite and deeply lobed, clustered in a swirl at the base of the plant. They have a fernlike appearance. Yellow flowers, sometimes reddish yellow or lavender, appear in late spring, clustered in swirls at the tips of spikes. They are two-lipped and modified for pollination by strong bees. Fruits are capsules shaped like a flour scoop, with many small seeds.

A tea made from the plant is used as an herbal remedy.

Gratiola aurea Muhl.,
golden hedge-hyssop (Plantaginaceae)

One of the two species in the genus that occur in the Adirondacks, golden hedge-hyssop (see discussion in the next entry for the derivation of names) is a perennial that grows from rhizomes and produces stolons (runners) that extend from the plant to form an often dense stand. Plants grow to about a foot and a half tall and have opposite leaves that are usually covered with black dots and lack conspicuous teeth. Golden yellow flowers (*aurea* means "golden") appear in midsummer. They are tubular and up to three-quarters of an inch long, with two stamens.

It grows along the shores of lakes and streams.

Gratiola neglecta Torr., northern clammy
hedge-hyssop (Plantaginaceae)

This is a modest, easily unnoticed plant (hence the specific epithet, *neglecta*, meaning "overlooked"). The genus name is thought to be a corruption of *gratia Dei* ("by the grace of God") because of the medicinal value of several European species long used for a variety of cures. It is called "hyssop" in allusion to the purported taste and efficacy of a European species likened to the biblical hyssop. While the "clammy" descriptor is clearly derived from the sticky hairs of the plant, the reason that it is called "hedge-" hyssop remains a mystery.

A rapidly growing annual up to about a foot tall with opposite, crowded leaves on single or loosely branched stems, it produces honey-colored, tubular flowers about a quarter of an inch long with five incised petals. Like some other members of the plantain family, the Plantaginaceae, northern clammy hedge-hyssop has only two stamens.

This is a wetland plant found in swamps, ditches, muddy ruts, and areas that were flooded and draw down during the summer. It flowers throughout the growing season.

Linaria vulgaris Mill.,
butter-and-eggs (Plantaginaceae)

The colorful flowers of this perennial are the reason the plant was introduced from the Mediterranean region to the United States as an ornamental. A common and colorful weed, it favors dry fields, roadsides, and other open disturbed areas. Plants grow to about three feet tall and have narrow, alternate leaves. Flowers, often borne throughout the growing season, are two-lipped, with a pale yellow lower lip and a dark yellow upper lip and resemble a miniature snapdragon (to which butter-and-eggs is related). The fruit is a capsule filled with flattened seeds that are carried by air currents.

Butter-and-eggs is reported to be edible and has been used in herbal remedies, especially in Europe, where it is also common.

Lysimachia terrestris (L.) Britton, Sterns, & Poggenb., swamp candles (Primulaceae)

"Swamp candles" is an appropriate name for this widespread inhabitant of shores, marshes, and forested lowlands because of its bright yellow, cylindrical, dense masses of flowers in the summer. Flowers are star-shaped, with five petals and a red center. Petals are often streaked with black resin canals. The flowers are unusual in producing oil rather than nectar to attract the melittid bees that are pollinators.

Stems, usually single but sometimes branched, of this perennial arise from stolons. Leaves are opposite and tapered at both ends. Late in the season, bulblets resembling reddish caterpillars develop in the axils of the leaves. As fall approaches, these fall to the ground, where they can be carried by water to new sites. The plant also reproduces by seed as well as stolons.

Lysimachia thyrsiflora L.,
tufted loosestrife (Primulaceae)

Like other loosestrifes, this species has bright yellow flowers but is distinguished from the other four yellow-flowered species by the globe-shaped groups of flowers produced only on lateral branches. It also is unique in having deeply dissected petals, giving the appearance of having more than its usual five. Tufted loosestrife is part of the worldwide northern guild of plants at home in the Adirondacks, where it grows in wetlands such as swamps and bogs.

Plants are about three feet tall and have opposite, elongate leaves about five inches long with a conspicuous midrib. Flowers appear in spring and are followed by small capsules with unusual triangular seeds. As with other loosestrifes, this species may have numerous purple dots on the leaves and petals.

The specific epithet, *thyrsiflora*, refers to the pattern in which the flowers develop—namely, a "thyrse," in which the main axis of the inflorescence continues to grow and produce additional flowers (indeterminate growth) while the side branches produce a set number of flowers (determinate growth).

Caltha palustris L., marsh marigold (Ranunculaceae)

Marsh marigold is a true harbinger of spring. The landscape of brown grasses and sedges brightens with marsh marigold's lemon-yellow flowers. These appear at the same time as pussy willow and buckbean flower, also denizens of wet areas. Plants are close to the ground, seldom more than two feet tall.

Leaves arise from the base of the plant, are round with a heart-shaped base, and have coarse teeth. Each plant can produce many flowering stems, each with a single flower the diameter of a half-dollar, containing five petals. The fruits, dry follicles that open along a slit, yield small, brown seeds.

The origin of the common name is unsettled but is apparently derived from the European use of a similar-looking but

unrelated plant that decorated churches for St. Mary. Marsh
marigold is found throughout the northern latitudes and is also
widespread in Europe and across much of North America.

Caution is advised in using this species for food, though it is
often included as an edible for its young leaves and young flower
buds. The toxin it contains is reported to be heat labile, so the
plant should never be eaten raw.

Ranunculus acris L., tall buttercup (Ranunculaceae)

An alternative name is "bitter buttercup" because of the bitter
(the specific epithet *acris* means "acrid" or "bitter"), toxic compound, protoanemonin, that has caused poisoning in cattle, especially cows. The genus name comes from *ranunculus*, Latin
for "little frog," because of this buttercup species' affinity for
wetlands in which frogs occur.

Tall buttercup is an introduction from Europe thought to have been brought as an ornamental garden plant that has now escaped to much of the United States and Canada. It is a fibrous-rooted perennial, growing three feet tall with alternate, deeply palmately lobed leaves (that is, with the lobes radiating from a central point), usually cleft in three to five sections. Flowers are bright yellow with five petals. Fruits are a cluster of small seeds with curved beaks, likely a mechanism for animal dispersal.

Several other buttercups occur in the Adirondacks. The kidney-leaved buttercup, *Ranunculus abortivus* L., is a native plant that is often abundant in rich, moist to wet fields and forests, and disturbed sites. Basal leaves are kidney-shaped, as the common name suggests. Flowers are about one-quarter of an inch in diameter, with five yellow petals that are so reduced in size that they appear "aborted" (hence the species name).

The bulbous buttercup, *Ranunculus bulbosus* L., arises from a bulblike structure; it is also an introduced species. The lower leaves of bulbous buttercup usually have five leaflets, and the plant grows in open, disturbed areas.

Northern swamp buttercup, *Ranunculus caricetorum* Greene, grows in swamps, fens, and lake margins. The specific epithet is descriptive—it means growing among sedges (species of *Carex*). A frequently used taxonomic synonym is *Ranunculus hispidus* Michx.

Another native species with features like those of the above is the hooked buttercup, *Ranunculus recurvatus* Poir., which grows in moderately moist forests.

Thalictrum dioicum L.,
early meadow rue (Ranunculaceae)

Early botanists coined the common name "meadow rue" because of the purported resemblance of the leaves of this plant
to those of rue (*Ruta graveolens*), at the time a well-known
European herb. But there is no relation between this tall plant
of mesic (moderately moist) woods and forest margins and the
European herb.

Early meadow rue is shorter than tall meadow rue, also
common in the Adirondacks, and does not grow in wet soil. It
also flowers earlier, along with other spring ephemerals. Leaves
are large, bi-compound, and arranged alternately on the stem.

Plants are strictly unisexual, bearing either only male
(staminate) or female (pistillate) flowers. The pistillate flower plants are less conspicuous because their flower-size is
reduced to a minimum, as the plants are wind-pollinated
and have no need to attract animal pollinators. Male plants

are more obvious because of the large number of flowering stems, each with many staminate flowers, with ten dangling stamens apiece.

Geum fragarioides (Michx.) Smed., barren strawberry (Rosaceae)

When in fruit, this would never be confused with a true strawberry (species of *Fragaria*) because it does not produce a fleshy, red fruiting structure. Furthermore, the barren strawberry has yellow flowers produced in the spring, whereas strawberry has white flowers. Its fruits are dry and hard. So why is it called a strawberry?

The leaves bear a close resemblance to the leaves of the two Adirondack species of strawberry. However, barren strawberry leaves are evergreen. Unlike true strawberries, this species does not produce above-ground stolons. But it does spread extensively via rhizomes, often forming dense patches. It prefers open patches in forests, especially on thin and rocky soil, and is found in dry conifer forests.

The genus is listed as *Waldsteinia fragarioides* (Michx.) Smelmak in most guides.

Geum rivale L., purple avens (Rosaceae)

Some plants of this species may have yellow flowers as well as various shades of red and purple. A plant of damp, wet areas influenced by calcium, in full shade or open fens, it is easy to overlook the flowers, which are borne about knee high. Flowers are five-parted, with the petals often reflexed in mature flowers. The individual fruits are small, with a hooklike structure, an adaptation for animal dispersal. Only a few leaves are on the stem when flowering; most are basal and lyrate-shaped leaves (with several lobes, the smallest at the base).

One of the lesser-known common names for this widespread northern species is "chocolate root," because the root after boiling can be used as a chocolate substitute. The root was formerly used in making beer.

Potentilla norvegica L., ternate-leaved cinquefoil (Rosaceae)

The uninspiring common name, "ternate-leaved," refers to the leaf shape. "Ternate" means arranged in threes. So this is the cinquefoil with three-parted leaves, rather than the usual five leaflets of most species of *Potentilla*. The full common name very confusingly then means "the three-leaved five leaf." *Norvegica*, the specific epithet, means "from Norway," a reminder of the very broad distribution of this widespread species that is native to New York.

Flowers are yellow, five-parted, and narrow to the base, where they attach to the receptacle. This gives the appearance of widely spaced petals. The fruit is a cluster of dry, hard seeds. This cinquefoil most frequently grows in dry, open, disturbed sites. A tall, coarse plant, it can reach a height of two feet.

Potentilla recta L., sulphur cinquefoil (Rosaceae)

This is a true cinquefoil (French for "five leaves") because the alternate compound leaves have five leaflets (some upper stem leaves may have only three). The leaves bear a close resemblance to those of marijuana (*Cannabis sativa*) but are readily distinguished by the lack of the characteristic fragrance of marijuana, hairiness, and the broader teeth of the cinquefoil. Despite these differences, sulphur cinquefoil has been collected and sold as marijuana, even though it does not contain the drug plant's alkaloids.

The entire plant is hairy, at least early in the season. A single stem that grows up to two and a half feet bears bright yellow flowers. These have five petals, numerous stamens, and a mound of pistils (ovary-containing structures) in the center. Each petal has a notch at the tip.

An introduction from Europe, sulphur cinquefoil prefers open, sunny, often disturbed areas.

Verbascum thapsus L., common mullein (Scrophulariaceae)

In Germany, this common, weedy, but attractive import from Eurasia is often known as "king's candle," an apt image for this tall (up to eight feet), robust biennial. Attractive yellow flowers appear from mid-summer until frost. Flowers are crowded at the tip of the plant, but only a few are open at one time. They have five petals and conspicuous, hairy stamens. Common mullein grows in open, disturbed areas, favoring railroad embankments, roadsides, and rocky sites.

During the first year of growth, the plant forms a rosette of large, densely hairy leaves. The structure of the hairs is complex. They are either star-shaped or dendritic (structured like a tree) and give the leaves a distinctly woolly appearance. A 10× hand lens will reveal the intricate structure and beauty of these hairs. While most animals avoid eating the woolly leaves, the flowers draw bees and other insects. An unusual group of bees, carder bees, collect the hairs to line their nests.

The plant is used in compounding herbal remedies, especially in Europe. Mullein should not be ingested, however, as there are reports of toxic substances in the plant.

Viola pubescens Aiton, downy yellow-stemmed violet (Violaceae)

The common name accurately describes this widespread violet—it is hairy (pubescent). The clear yellow flowers have purple stripes at the mouth of the corolla. Like other violets, these lines are likely guides for floral visitors. The entire plant is covered with long, soft hairs. Cleistogamous (hidden) flowers are clustered on short stems near the base of the plant. Flowering in midspring, this violet is found in mesic (moderately moist) upland forests.

Viola rotundifolia Michx.,
early yellow violet (Violaceae)

The most distinctive violet in terms of its growth pattern: fleshy, round leaves that lie flat on the ground. A yellow-flowered, stemless violet, it is one of the earliest to flower. This species favors mesic forests and streambanks. It is often common along paths.

Red Flowers

Asarum canadense L., wild ginger (Aristolochiaceae)

This is no relation to true ginger, a tropical plant in the family of the same name (Zingiberaceae). But it does have an aroma like ginger. Unlike true ginger [*Zingiber officinale* (L.) H. Karst.], wild ginger is toxic because of the presence of aristolochic acid and should never be ingested—this despite numerous recommendations for using it in wild food recipes.

An attractive creeping plant only a few inches high with heart-shaped leaves, wild ginger produces reddish flowers in the early spring that are easily overlooked because they are borne directly on the soil surface in rich woods, generally over limestone. The flowers have a distinct adaptation for beetle pollination, what has been described as "mess and scatter." The animal enters the flower, drawn by the pollen and other flower parts. While crawling around the flower, the beetle transfers

pollen. But most reports indicate that the flowers are self-pollinating. The proverbial further research is needed.

The fruits have to be carefully searched for; they are greenish and contain about a dozen brown seeds with an oily, white,
fleshy appendage, an elaiosome, which attracts ants that harvest and scatter the seeds.

Lobelia cardinalis L., cardinal flower (Campanulaceae)

The common, widespread cardinal flower is unique in several ways. First, it has the most
spectacular display of bright-
red flowers of any plant in the
Adirondack flora. It is also unusual in being adapted to one
pollinator, the ruby-throated
hummingbird, which is at-
tracted to red flowers.

Its floral mechanics are remarkable. The flowers of this wetland plant are borne on stems up to five feet tall with alternate
simple leaves. Nectar is at the bottom of the corolla tube, which
is the reward for the hummingbird's visit. But the plant has
clever devices to benefit from its nectar production. The flower
is functionally male at first, when a delicate brush appears at
the tip of the floral tube, where it deposits pollen onto the head
of the visiting hummingbird. About five days later, the brush is
replaced by the expanding stigma, which will touch the bird's

head to receive pollen, ensuring cross-pollination. But should no hummingbird show up, the style will bend down to pick up its own pollen.

Cardinal flowers are frequent garden subjects that provide a unique scarlet display in midsummer. Some varieties have purple-tinged leaves that enhance an already spectacular plant. By early fall, the capsules open to release the abundant, tiny seeds, which need a saturated but bare surface upon which to germinate and become established.

Monarda didyma L., Oswego tea (Lamiaceae)

Oswego tea is less common in the Adirondacks than the closely related wild bergamot (*Monarda fistulosa*) and grows in different habitats. It favors forested seeps, streambanks, and moist roadsides and is often a local escapee from gardens, where it is a valued subject because of its ease of culture and striking beauty.

Oswego tea resembles its congener in growth habit, with opposite leaves that are green, sometimes gray green. Flowers are of the same morphology as those of wild bergamot but are vermillion to bright scarlet in color. Pollinators include hummingbirds and bees.

Aquilegia canadensis L., wild columbine (Ranunculaceae)

Few of our wildflowers have flowers as distinctive as the columbine. This spring-flowering perennial of usually open woodlands grows to a height of about four feet and has alternate compound leaves composed of three leaflets. Flowers are eye-catching in both shape and color. The five petals are yellow and are subtended by highly specialized modified sepals about an inch long, terminating in a knoblike reservoir of sweet nectar. Sepal tips are arranged in a circle like the points on a crown. The high-energy nectar attracts hummingbirds and long-tongued insects. The fruits begin developing in late spring, and by summer the five follicles open to release seeds.

Columbine is a good garden subject because of its perennial nature and because it is easy to grow from seed as well. Since the plant readily self-pollinates, large numbers of seeds are produced, ensuring that a patch of columbine will firmly establish itself in the garden.

Comarum palustre L., marsh cinquefoil (Rosaceae)

 Formerly known as *Potentilla palustris*, the marsh cinquefoil is the only truly aquatic cinquefoil in the Adirondacks, usually occurring in pockets of standing water in emergent marshes, along water courses, and in roadside ditches that are permanently saturated. In midsummer, one to a few flowers about three-quarters of an inch in diameter are produced at the end of a stalk. These flowers consist of five purplish-red petals and larger, brownish-red sepals. Leaves are grayish-green, pinnately compound, generally consisting of five (sometimes seven) toothed, oblong leaflets, but fewer further along the stem up to the flowers. Stems are stout, often lying on mud and rooting at their nodes. The fruit resembles a small, dried strawberry. Plants can reach about two feet tall.

Sarracenia purpurea L., purple pitcherplant (Sarraceniaceae)

The Adirondack flora has three genera of insectivorous—or more precisely, carnivorous—plants: bladderworts (species of *Utricularia*), sundews (species of *Drosera*), and a single pitcherplant, the purple pitcherplant. It is unusual among its meat-eating fellows as the only one with a passive trap. Insects are lured into the pitcher, captured, and digested. The leaves are up to a foot long and are recumbent on the soil. Attractants around the lip of the pitcher draw the prey, which then are entangled in downward-pointing hairs just above a slippery vertical section that contains the liquid filling the bottom of the pitcher. Once the insect slides down the pitcher wall and is in the liquid, its fate is settled.

Purple pitcherplant is often abundant in Adirondack bogs, as well as at the peaty margins of streams and lakes. These are areas with poor soil nutrition, especially nitrogen. So the plant augments what is lacking by deriving nitrogen from its carnivory.

Pitcherplants grow close to the ground but send up flowering spikes, each with a single flower. The morphology of the flower is as remarkable as the plant's nutrition. Flowers are maroon, with five drooping petals. The large, umbrella-shaped structure in the middle of the flower is the style. No other plant has a similar style. In the majority of angiosperms, the style is long and narrow, like a stylus or pencil. A five-parted capsule develops in the summer with numerous round, brown, hard seeds. The plant is easy to grow from seeds after they have been cold-treated and scarified.

Pink to Purple Flowers

Asclepias incarnata L., swamp milkweed (Apocynaceae)

This milkweed is distinguished by its long, pointed opposite leaves that lack teeth on the margin. Plants may be five feet tall. It is the only milkweed in the Adirondack flora that is a wetland plant, found in marshes, borders of streams and lakes, and other open wet areas.

Light pink flowers, each about a quarter of an inch across, are produced in flat-topped clusters in mid- to late summer and have a pleasant fragrance likened to cinnamon. As with other milkweeds, this is a food plant for the larvae of monarch butterflies and is now being widely planted to feed this threatened insect. Although a food plant for the butterflies, this milkweed, like others in the genus, is toxic to humans because of the presence of cardiac glycosides that can interfere with regulation of heartbeat.

In the fall, narrow, erect pods are produced. These split open to release the seeds, each of which has a tuft of silky hairs to ensure that the seed is airborne.

Asclepias syriaca L., common milkweed (Apocynaceae)

This is the most abundant milkweed of the region, a group of perennials with usually opposite leaves and milky juice. The flowers are complex and highly specialized for bee pollination, in that the leg of the bee engages a strap that is attached to two small bags of pollen. When visiting the next flower, the bee drops the pollen mass onto the receptive stigma, rather like a key fitting into a lock. The fruit of milkweeds is a follicle, often referred to as a pod, that splits open to release the seeds, each of which has long, silky hairs to ensure dispersal by wind. These silky hairs have been used in the production of flotation devices and were harvested from the wild during the Second World War.

Why is the most common milkweed—a native plant—
named after Syria? Ascribing plants to areas where they did
not grow was not an uncommon phenomenon during the era of
botanical exploration. The piles of dried specimens sometimes
were mixed, as in the case of the naming of this plant.

In recent years, common milkweed has drawn a lot of attention because it is a host plant of the monarch butterfly. It is
easy to grow and will spread considerably by rhizomes. Since
the butterfly larvae prefer young, tender leaves, older stems can
be removed to stimulate new stem development.

Symplocarpus foetidus (L.) Salisb. ex W.P.C. Barton,
skunk cabbage (Araceae)

Skunk cabbage is unique. It is the first of our native plants to
flower and has the biggest leaf of any of our wetland flora. It is
one of a guild of plants displaying thermogenesis, that is, the

ability to produce heat, and it is therefore not uncommon to see the flowering structure melt the snow around it.

Like Jack-in-the-pulpit, skunk cabbage has a specialized floral structure consisting of a thickened axis, the spadix, wrapped in a covering called the spathe. The spathe is about six inches long, reddish brown and often striped, pointed at the top, and with a longitudinal slit. The unisexual flowers are embedded in the fleshy spadix and are insect pollinated. Aerodynamic studies show that the structure of the spathe and its slit create a vortex of air around the spadix that can effect pollination. This mechanism is a back-up if the flies and other carrion eaters drawn by the foul volatiles produced by the heated flower are not available.

All parts of the plant contain compounds akin to the scent of skunks. The leaves appear after flowering and can be three feet long and about as broad. They arise from an erect rhizome. Skunk cabbage has contractile roots that pull the plant into the soil, perhaps an adaptation to growth in a wetland environment when there is an accretion of soil. (Contractile roots are widespread in plants, for example, dandelion.) Skunk cabbage prefers areas that are permanently wet with groundwater discharge.

The plants are important food for black bears. In some regions, the diet of bears consists of more than 50% skunk cabbage, especially in spring. Bears must be able to deal with the large quantities of calcium oxalate in the leaves and rhizomes. Humans cannot, so ingesting skunk cabbage should be avoided.

Centaurea stoebe L. ssp. *micranthos* (S. G. Gmel. ex Gugler) Hayek, spotted knapweed (Asteraceae)

Centaurea jacea

Centaurea nigra

Accidentally introduced from Europe in the late 1800s, spotted knapweed is now widely naturalized throughout the United States along roadsides and in old fields, waste places, and other disturbed areas. The common name is derived from the Old English word *knop* ("knob") because of the hard nature of the flowering head. It is considered a serious invasive species in most places. Flower heads are one inch in diameter and consist only of pink to purple disk flowers that have a starry appearance from early summer until fall. Bracts beneath the flower

head are black-tipped. Leaves are grayish blue-green and pinnately divided with narrow lobes. Plants reach about three feet
in height.

   Brown knapweed, *Centaurea jacea* L., and black knapweed,
Centaurea nigra L., referring to the colors of the bracts of the
flowering head, also occur in the Adirondacks on similar sites,
and have unlobed or shallowly lobed leaves. None of the knap-
weeds in New York are native.

Cirsium arvense (L.) Scop., Canada thistle, creeping thistle (Asteraceae)

Cirsium muticum   *Cirsium vulgare*

Canada thistle is from Europe but is widely naturalized in the
eastern United States in old fields, waste places, and other disturbed areas, and along roadsides. It is a serious agricultural
weed and difficult to eradicate owing to its extensive and deep
root system. The flower head is one-half of an inch wide and
consists of pale purple disk flowers. Many heads are arranged
in flat-topped clusters from summer until fall. Leaves are spiny,
pinnately divided, and whitish below. Stems are also spiny but
hairless. Plants reach about four feet in height.

   The bull thistle, *Cirsium vulgare* (Savi) Ten., is another
non-native thistle of the Adirondacks that occurs in habitats

similar to those that foster the Canada thistle. It is also considered a noxious weed throughout the United States. Its leaves are very spiny and not whitish below; flower heads are up to one and one-half inches wide. Stems are spiny-winged. Plants produce a rosette of leaves the first year, followed in the second year by a flowering stem that reaches six feet in height.

The native swamp thistle, *Cirsium muticum* Michx., occurs in the Adirondacks in swamps, marshes, and fens. Flower heads are about one and one-half inches wide and consist of purple to pink disk flowers. Leaves are deeply pinnately lobed; basal leaves are large, reaching sixteen inches in length and six inches in width. The swamp thistle reaches about seven feet in height. Flowers of swamp thistle attract many insects, especially butterflies and long-tongue bees. Small songbirds like American goldfinch eat its seeds.

Eutrochium fistulosum (Barratt) E. E. Lamont,
hollow Joe Pye weed (Asteraceae)

The hollow Joe Pye weed occurs in open wetlands, especially meadows, marshes, and streamsides, reaching nine feet tall. It flowers in mid- to late summer, its pink disk flowers in round clusters that attract the spicebush swallowtail, other butterflies, and bee species. Leaves are in whorls of four to seven.

The descriptor "hollow" refers to the fact that its purple, waxy stem generally is hollow, in contrast to the usually solid stem of spotted Joe Pye weed and purple-node Joe Pye weed.

Eutrochium maculatum (L.) E. E. Lamont, spotted Joe Pye weed (Asteraceae)

Spotted Joe Pye weed is common in open wetlands, especially shallow emergent marshes and streamsides, reaching a height of seven feet. Flower heads have from nine to more than twenty pink to purple disk flowers in a flat or round cluster, in mid- to late summer. Flowers are pollinated by various species of butterflies and bees. Leaves are usually in whorls of four or five, not exceeding five.

This species is similar to hollow Joe Pye weed but has spotted stems (*maculatum*, Latin for "spotted").

Eutrochium purpureum (L.) E. E. Lamont, purple-node Joe Pye weed (Asteraceae)

Purple-node Joe Pye weed is similar in appearance to these other species but often occurs in drier, forested habitats, and its stem is hard, mostly solid, and purple at the nodes.

Oclemena nemoralis (Ait.) Greene, bog aster (S3) (Asteraceae)

The bog aster is confined to open fens and bogs and, because of its rarity, is state-protected. Flower heads are one to one and one-half inches in diameter and consist of about thirteen to twenty-seven purplish-pink ray flowers that surround bright yellow disk flowers. Heads are usually solitary or few in a flat-topped cluster. Leaves are narrow, hairy below; margins are rolled under (revolute).

Symphyotrichum novae-angliae (L.) G. L. Nesom, New England aster (Asteraceae)

Symphyotrichum novi-belgii

Arguably our most beautiful native aster, with violet-purple ray flowers around a center of yellow disk flowers in a head the size of a half-dollar. Frequent throughout the Adirondacks along roads and in old fields, it grows up to six feet tall with leaves that clasp the roughly hairy stem. This is one of our few native asters that is widely cultivated.

The New York aster, *Symphyotrichum novi-belgii* (L.) G. L. Nesom, is also widespread in the Adirondacks. The common name "New York aster" does not appear to have much to do with Belgium (*belgii*), but New Belgium was an early name of New York. Its appearance is much different from that of the New

England aster, with narrow, non-clasping leaves; smaller flower heads; and a preference for wetter areas.

Symphyotrichum prenanthoides (Muhl. ex Willd.) G.L. Nesom, crooked-stemmed aster (Asteraceae)

The crooked-stemmed aster occurs in open to semi-open, wet conditions, including wet ditches, swamps, and streambanks. Flower heads are about one inch in diameter and consist of twenty to thirty-five pale blue to lavender ray flowers surrounding the central yellow disk flowers. It is very similar in appearance to *S. novae-angliae* except that the leaves of the crooked-stemmed aster have a winged stalk that clasps the stem. Plants reach about forty inches in height and often grow in patches, from rhizomes.

Symphyotrichum puniceum (L.) Á. Löve and D. Löve, purple-stemmed aster (Asteraceae)

The purple-stemmed aster occurs in wet meadows, marshes, fens, and lowland forests and along streambanks. Flower heads are about one inch in diameter and consist of thirty to sixty lavender to bluish ray flowers that surround the central disk flowers. It blooms in late summer into fall. Leaves are somewhat shiny and narrow, with basal lobes that clasp the stem, which turns reddish-purple as the plant matures. Plants can reach seven feet tall and are bristly white hairy.

Hesperis matronalis L., dame's rocket (Brassicaceae)

This plant has nothing to do with the space program. Rather, it is called "rocket" after the French *roquette*, the name for arugula (*Eruca sativa*). Both are in the mustard family (Brassicaceae). Many who see this plant from a distance readily misidentify it as a *Phlox*. But on close inspection, one can count the four petals in a cross-shaped arrangement, typical of the mustard family, versus the five-petaled flowers of *Phlox*.

Dame's rocket is an attractive species naturalized from Eurasia, and it flowers in late spring. It is often included in "wildflower" seed mixes. Although it is indeed a wildflower, it is important to recognize that this is not a native species, despite its intentional seeding along roadsides. Its purple, pink, or white flowers are fragrant at night, the origin of the genus name *hesperis*, after the evening star. It is a mildly invasive plant, growing to five feet tall. Old homesteads, roadsides, and open woods are frequent habitats.

Linnaea borealis L., twinflower (Caprifoliaceae)

Part of a guild of plants requiring "belly botany," which requires you to get on the ground to study and appreciate them, twinflower is abundant in the Adirondacks, growing in dry and moist upland forests, often on decaying logs as well as hummocks in bogs and fens.

Flowers are produced in pairs at the top of a flowering stem about three inches tall. The tiny, fragrant flowers are bell-shaped, pink or white and often striped with purple. Twinflower is a creeping vine with small, oval evergreen leaves with hairy margins. Without flowers, it can be confused with snowberry [*Gaultheria hispidula* (L.) Muhl.], which has opposite, round leaves lacking abundant hairs on the margins. Both twinflower and snowberry can grow in the same habitat.

Twinflower was a favorite of the "father of botany," Carl Linnaeus (1707–78), who made it well-known to Europeans. It has been featured on Swedish currency, and "Linnaea" is used as a girl's name in Sweden. Twinflower grows in northern regions around the globe.

Saponaria officinalis L., bouncing bet, soapwort (Caryophyllaceae)

Soap with pink flowers is one way to describe this hardy roadside perennial introduced from Eurasia. The entire plant, and especially the root, contains saponins—the same compounds that make soap foam when it is mixed with water. In fact, you can take a bunch of soapwort leaves and crush them under water to obtain a weak but effective soap.

Another common name is bouncing bet ("bet," from Elizabeth), which conveys the attractiveness of the large masses of pink flowers on the stem with opposite, dark-green leaves. The five petals, like those of most other members of this family, are notched at their tips and arise within a tube-shaped calyx. It flowers in midsummer and produces fruits in early fall. These are capsules with numerous brown seeds.

Saponaria officinalis was named by the great Swedish botanist, Linnaeus, based on the use of the plant for soap as well as medicine. The specific epithet *officinalis* refers to the use of the plant for medicine. Linnaeus used the same specific epithet for numerous plants that were "official," that is, known for their medical use.

Epigaea repens L., trailing arbutus (EV) (Ericaceae)

Trailing arbutus does well in margins of forests, edges of cliffs, and roadsides, and it favors other open areas with acidic soil. It often grows in patches with blueberries and huckleberries, other citizens of acidic soil.

Both the genus name (*Epigaea*, "upon the earth") and specific epithet (*repens*, "on the ground") describe the habit of this charming, early spring–flowering plant. Trailing arbutus is technically a "subshrub" because of the woody nature of its stem. The evergreen leaves are alternate, oval, and covered with rusty hairs, at least when young. They have a distinctive reticulate (netted) surface. Dense rusty hairs cover the stem. Flowers are five-lobed, trumpet-shaped, and varying shades of pink. The fragrance is usually described as spicy. Flowers are often partially hidden under the leaves.

Because of its beauty, it is a state-protected plant species. Widespread throughout the Adirondacks and beyond, trailing arbutus is part of the culture of eastern North America, as evidenced by the poem "The Trailing Arbutus," by John Greenleaf Whittier (1807–92), quoted in part here:

> I wandered lonely where the pine-trees made
> Against the bitter East their barricade,
> And, guided by its sweet
> Perfume, I found, within a narrow dell,
> The trailing spring flower tinted like a shell
> Amid dry leaves and mosses at my feet.

Pyrola asarifolia Michx., pink shinleaf (S2) (Ericaceae)

This is arguably our most beautiful shinleaf because of the pink or pink-tinted white flowers. It can also be distinguished from the other shinleafs by the shape of the basal leaves, which are heart- or kidney-shaped, with a very shiny surface. Its habitat is different as well, with pink shinleaf favoring moister sites such as low-land forests. It spreads extensively by rhizomes, often forming large patches. Because of its rarity (S2) in New York, it is state-protected.

Once placed in their own family, the Pyrolaceae, shinleafs are now in the same family as blueberries, the Ericaceae.

Amphicarpaea bracteata (L.) Fernald, hog peanut (Fabaceae)

Hog peanut is a vine that twines throughout adjacent vegetation, typically in shaded, moist conditions, including floodplain forests and other low, wooded landscapes. Its leaves are trifoliate, that is, consisting of three leaflets, and the plant resembles other members of the bean family. Its above-ground flowers in late summer are about one-half inch long, pale violet and white, tubular, and in clusters. They are readily recognized as being related to the flowers of beans, peas, and other members of this family—as is the fruit, which is a green pod (legume). Self-fertile flowers lack petals, occur at the base of the plant (or just beneath the surface), and produce a fleshy, edible fruit. This fruit should be cooked for best flavor. The genus name means "two kinds of fruit."

Apios americana Medik., groundnut (Fabaceae)

Groundnut, also known as potato bean, is a perennial that produces annual vines that can climb up to fifteen feet, supporting itself on surrounding vegetation. The leaves are about three inches long with three or five leaflets. Attractive purple-brown flowers are produced in midsummer, followed by narrow pods, with the flowers and pods revealing it as a member of the bean or legume family.

This is one of the few native plants seriously considered for development as a commercial crop because of the small, nutritious tubers produced in abundance on the rhizomes. These are borne in a sequence like a rosary, with strings of tubers sometimes several feet long. When boiled or roasted, they taste like a potato.

Groundnut favors margins of rivers and lakes or areas that have wet soil that dries out in the summer. Because it is easy to grow from the tubers, groundnut is sometimes planted in gardens.

Desmodium canadense (L.) DC., showy tick trefoil (Fabaceae)

Hylodesmum glutinosum

The showy tick trefoil is the most widespread of the *Desmodium* species in the Adirondacks, occurring in old fields and along roadsides, rocky streambanks, and forest edges. Plants are downy, covered with soft hairs. Flowers resemble those of peas, are numerous, about one-half inch long, pink to purple, and densely packed along the branched stalk. The fruit is a three- to five-jointed pod that is very sticky, readily adhering to animals and clothing, as ticks do. Leaves are scattered along the stem in threes, hence the name "trefoil." Plants tend to grow in clumps and can reach six feet in height, although typically they are about half that height. *Desmodium* is Greek for "bond" or "chain," referring to the jointed seed pods.

The pointed-leaved tick trefoil, *Hylodesmum glutinosum* (Muhl. ex Willd.) O. Ohashi and R. R. Mill, also occurs in the Adirondacks, in dry to moist hardwood forests. The leaves of this species are restricted to below the flowering cluster, in a whorl-like arrangement. Flowers are pink but are more scattered along the stalk. The fruit is two- to three-jointed. The genus of this species is listed as *Desmodium* in many guides. *Glutinosum* means "very sticky." All of the tick trefoil species are nitrogen fixers.

Geranium maculatum L., wild geranium (Geraniaceae)

The wild geranium occurs in open woodlands and along forest edges. Flowers are at least one inch in diameter, consisting of five pink to purple unnotched petals. Flowering occurs in mid- to late spring. The fruit is a long, erect, five-parted capsule, each part containing a long, narrow, single seed. Seeds are forcibly ejected from the capsule when mature. The leaves have five deeply cut lobes. Plants often grow in clumps and can reach about two feet in height.

This geranium family is also referred to as the cranesbill family because of the long, beaklike ovary.

Geranium robertianum L., herb Robert (Geraniaceae)

The nativeness of herb Robert is in question; it is thought to have been introduced from Eurasia, though some botanists consider it native. In any event, herb Robert has become widely naturalized in damp rich woods, streambanks, moist waste places, and other disturbed substrates. It especially thrives on calcium-rich substrates. Flowers are two per stalk, about one-half inch in diameter, pink to reddish-purple, with unnotched petals and an unpleasant smell. The leaves are deeply lobed, with each lobe much narrower than those of the wild geranium.

The genus name is from the Greek, *geranos* ("crane"), referring to the narrow, beaklike fruit. A few other species of *Geranium* also occur in the Adirondacks and are typically distinguished by whether the petal margins are entire or notched. The perennial species of *Geranium* in New York are related to the horticultural plants known as geranium (genus *Pelargonium*), which often have many more flowers per stalk and often brighter colors, as well as unusual leaf lobing and color patterns.

Hypericum virginicum L., Virginia marsh St. John's wort (Hypericaceae)

The extensive marshes in the Adirondacks present a diversity of interesting plants. Some are easily overlooked unless they are in flower, including the marsh St. John's wort, a frequent plant of marshes, lake shores, and other open wet areas. It is usually a single-stemmed perennial with opposite leaves lacking a leaf stalk; the leaves often wrap about the stem, which grows up to four feet tall. Its attractive mauve flowers display three groups of stamens, each group a cluster of three. The fruit is a pointed, many-seeded capsule, usually with three persistent styles.

A closely related species, Fraser's marsh St. John's wort, *Hypericum fraseri* (Spach) Steud., is less common but is found in habitats similar to those that support *H. virginicum*. Fraser's marsh St. John's wort is distinguished by the rounded top of its capsule.

Monarda fistulosa L., wild bergamot (Lamiaceae)

Aromatic oils are so abundant in this species that on a hot summer afternoon the air is filled with the distinctly bergamot fragrance. This rigorous perennial grows in open, sunny, often infertile areas. Leaves are gray-green and opposite, produced on the square stems characteristic of the mint family. Flowers are abruptly curved, lavender, and two-lipped. They are produced in a circular pattern on a cluster of modified leaves (bracts) that are usually pink-tinged. Flower visitors include hummingbird moths, bumblebees, and butterflies.

The entire plant is suffused with a cocktail of aromatic compounds. The dried leaves are used to make a tea. Soaking a handful of fresh leaves in vodka for two months makes a tasty aperitif.

Streptopus lanceolatus (Aiton) Reveal, rose twisted stalk (Liliaceae)

The rose twisted stalk occurs in moderately moist hardwood and mixed hardwood conifer forests. Its tubular flowers on slender, bent stalks are pink, streaked with dark red, and occur singly along the stem below each leaf node. The fruit of this species is a round, reddish-purple to red berry. Leaves are lancelike and stalk-less, but they do not clasp the branched, zigzag stem. Its overall stature is similar to that of the clasping-leaved twisted stalk. The species name for quite some time was *Streptopus roseus*.

Decodon verticillatus (L.) Elliott,
water willow (Lythraceae)

Water willow is unfortunately also known as "swamp loose-strife." With this other common name and its bright pink flow-ers, the native water willow is sometimes confused with the highly invasive and non-native purple loosestrife. Water willow occurs in shallow, stagnant, or slow-moving waters of ponds, lakes, and streams, where its many long and slender branches create quite a tangle.

Flowers consist of four to seven crinkled petals and are located in widely spaced clusters at the leaf nodes along the branches rather than in a densely packed spike as are those of purple loosestrife. The fruit is a round capsule. Leaves are el-liptical to lancelike and occur in whorls or, less often, opposite one another. Although stem lengths can reach eight feet, total plant height is typically about half that length because of the arching nature of the stems, which bend and root in the water to form new plants.

Many insects pollinate the flowers of water willow. Its capsules are eaten by various duck species, and muskrats feed on its spongy stems.

Lythrum salicaria L., purple loosestrife (Lythraceae)

This very aggressive aquatic invasive has wreaked havoc in wetlands in the Adirondacks and throughout most of the United States and adjoining provinces of Canada. Purple loosestrife can be found in a diversity of wetland habitats in the Adirondacks but is most aggressive and abundant in disturbed sites. Leaves are long and narrow, and lack teeth. They resemble the leaves of some willows, species of *Salix*, hence the specific epithet. The stem is often densely hairy and grows five feet tall.

Beauty is the beast with this plant because it produces masses of attractive pink flowers while displacing native plants and reducing their food value for wildlife. The attractive flowers have encouraged the plant's entry into the horticultural trade. They are five-parted, purple, and like other members of this family, are crinkly when emerging from the bud. Plants sold for gardens were claimed to be sterile, but studies have shown that they could interbreed with wild plants. Three types of flowers are produced: long-styled, medium-styled, and short-styled. The length of the stamens inversely corresponds: long for short-styled, medium for medium-styled, short for long-styled. Only flowers from different types are fertile. Flowers can be successfully pollinated only by pollen from a different flower. This is unquestionably a successful strategy, as a single plant can produce a million seeds.

Claytonia caroliniana Michx.,
Carolina spring beauty (Montiaceae)

Named after the pioneer Virginia botanist, John Clayton (1694–1773), spring beauty is a charming and well-known spring ephemeral, a group of plants that produce leaves and flowers

Claytonia virginica

before tree leaves appear in the deciduous forests where they grow, after which the plant slowly senesces and disappears until the following spring. Of the two species of spring beauty in the Adirondacks, Carolina spring beauty, *Claytonia caroliniana* is more common than the eastern spring beauty, *C. virginica* L., and it prefers moister habitats than its congener. The two species, both typically less than five inches tall, are easily distinguished by the width of their fleshy leaves and their flowers. Leaves of eastern spring beauty are narrow and grasslike, while those of Carolina spring beauty are broader. Flowers of *C. caroliniana* are usually more pink than those of the other species, although this is not the best characteristic by which to separate the two.

Both species produce tubers that can be the size of a large olive and filled with starch. These are a good source of energy and were an important food for Native Americans. Harvesting the tubers, however, destroys the population and should be avoided. The entire plant is edible.

Chamaenerion angustifolium (L.) Scop., fireweed (Onagraceae)

Few Adirondack wildflowers are as dramatic as fireweed, common along roadsides and other open areas, especially sites that have been burned—hence the common name "fireweed." A large stand of these four- to six-foot-tall plants topped with large pink flowers is striking. After the Mount St. Helen's explosion in 1980, fireweed was the first plant to colonize open ground on the volcanic waste, providing a summer show of acres of pink.

In the Adirondacks it flowers in midsummer, releasing its airborne seeds with their long, silky hairs in August. A tall flowering stem will display buds at the top, then open flowers, below which are the thin, upward-pointing young fruits, and lower still, the tangled masses of the recurved fruits and the cottony hairs of the seeds. Fireweed is found all across the northern United States and Canada to Alaska and also occurs in northern Europe into Siberia.

Young shoots of the plant are edible, and remnants were found in the stomach of the remarkably preserved Tollund Man when he was discovered in a peat bog near Silkeborg, Denmark,

in 1950, two thousand years after his death. The fireweed shoots
were part of his last meal. The Irish poet Seamus Heaney wrote
about Tollund Man and his final repast:

> In the flat country near by
> Where they dug him out,
> His last gruel of winter seeds
> Caked in his stomach.

Arethusa bulbosa L.,
dragon's mouth (S2, EV) (Orchidaceae)

Also commonly known by its
genus name, arethusa, the flowers of few plant species rival the
beauty of this wetland orchid
that is primarily rose magenta
with a golden-yellowish to whitish bearded tongue. Dragon's
mouth occurs in open peatlands
that are either acidic (bogs) or alkaline (fens), often arising
from *Sphagnum* hummocks or floating mats. Where it occurs,
flowering individuals can number from dozens to hundreds; a
thousand or more are known from some sites. Peak flowering is
late spring to early summer, ending when two occasional associates, the grass pink (*Calopogon tuberosus*) and rose pogonia
(*Pogonia ophioglossoides*), begin to bloom.

Bulbosa refers to the bulblike, pea-sized, pearly white corms
from which this plant—with a slender, short, and solitary

leaf—develops after flowering. These corms were once used to treat toothache, but given this species' legal protection and that it is truly rare, it should not be dug or picked for any purpose. The literature suggests that late spring frosts can dramatically reduce the number of flowering plants from hundreds one year to few the next.

Calopogon tuberosus (L.) BSP., grass pink (EV) (Orchidaceae)

The grass pink, with its very slender solitary leaf, is nearly invisible until it flowers in late June until late July. Then, three to fifteen bright to dark pink flowers emerge on stems up to two feet long in acidic, open peatlands (bogs) and saturated, acidic sands in meadows and ditches. It nearly always occurs in association with rose pogonia (*Pogonia ophioglossoides*), and in acidic bogs, its flowering is followed a couple of weeks later by the white fringed orchid (*Platanthera blephariglottis*). Both species disappear as the black spruce (*Picea mariana*) and tamarack (*Larix laricina*) develop into closed-canopy forest. The grass pink also occurs in calcareous fens, typically on hummocks, removed from the water table. Where it occurs, individuals can number in the many dozens to even thousands. One unusual habitat for this species is on saturated but not deeply flooded sands or peat, for example, on iron mine tailings or in the ground below harvested peat.

The genus name is derived from the Greek *kalos* ("beautiful") and *pogon* ("beard"), referring to the tufts of yellow-tipped hairs on the upper petal or lip; *tuberosus* refers to the rather large tuberous corms of this species.

Medium to large bumblebee species are successful pollinators because they are heavy enough to lower the bumblebee and lip of the flower onto the column, where pollen from previously visited grass pinks has been deposited.

Corallorhiza maculata Raf., early spotted coralroot (EV) (Orchidaceae)

The pinkish to purplish-brown, leafless stems of the spotted coralroot hold up to three dozen flowers with purple-spotted lips in early summer. It occurs as scattered individuals or in clumps of up to twelve plants in dry to moderately moist, deeply shaded woods—typically those with abundant sugar maple (*Acer saccharum*), American beech (*Fagus grandifolia*), and eastern hemlock (*Tsuga canadensis*). Two distinct varieties of spotted coralroot are likely to occur in the Adirondacks: *C. maculata* var.

occidentalis (Lindl.) Ames and *C. maculata* var. *maculata*. The former variety is known as the early spotted coralroot because it usually blooms about two weeks earlier than the latter variety. A more dependable distinction between these varieties is that var. *occidentalis* has a lip that is broadened, versus a relatively narrow lip for *maculata*.

The genus name comes from *coral* ("coral-like") and *rhiza* ("root"), referring to the underground rhizomes that have a strong resemblance to marine corals; *maculata* means spotted, referring to the lip. Coralroots lack leaves and chlorophyll; because of their non-photosynthetic nature and dependency on fungi for nutrition, coralroot species do not emerge aboveground every year. In fact, these species are known to remain alive but belowground for years. The environmental trigger to produce flowering stems is not understood.

Cypripedium acaule Aiton, pink lady's slipper (EV) (Orchidaceae)

The most common lady's slipper species in the Adirondacks, the pink lady's slipper (also known as the pink moccasin flower) occurs on acidic sites, typically dry and shaded, and occasionally in bogs on elevated microtopography. This species is also frequent in pine plantations and natural stands dominated by eastern white and red pines (*Pinus strobus* and *P. resinosa*, respectively). In flower, it is one of the most conspicuous of Adirondack wildflowers in late May to early June. Like many of the native, terrestrial orchids with pink flowers (*Arethusa*, *Calopogon*, and *Pogonia*), individuals with white flowers are not uncommon. These individuals should not be confused with the small white lady's slipper (*C. candidum*) that is a wet-prairie and open rich fen species further south in New York, with a much smaller pouch and with leaves along the stem.

Pedium means "shoe," and *acaule* means "stemless"; flowering plants are not actually stemless, but the flowering stem lacks the leaves along the stems found in other lady's slipper species; this is the only lady's slipper species in the eastern United States lacking leaves along the flowering stem. Flowering individuals have two deeply and longitudinally ribbed basal leaves; single leaves are common in these populations. Without the deep ribbing, the leaves can be mistaken for a common associate, the blue bead lily (*Clintonia borealis*), or perhaps for other native orchids with two basal leaves. This species is rarely solitary and more commonly occurs in populations of dozens to many hundreds. In larger populations of pink lady's slipper, in which pollination by bumblebees is more likely to be successful, it is more frequent to find an occasional, remnant ribbed fruit capsule that in summer contains thousands of "dust seeds" (characteristic of the orchid family). *Cypripedium* species take ten to fifteen years or more to flower once seeds germinate. Its relative commonness among native orchid species suggests that it would be easy to grow in gardens—nothing could be further from the truth. Like all orchids native to New York, it is illegal to dig this species or pick its flowers on state-owned lands or on any private lands without permission of the landowner. So even if the putative medicinal properties (as a sedative and antispasmodic) of *Cypripedium* species were proven, one should not dig this or any specimen of this genus.

Cypripedium arietinum R. Br., ram's head lady's-slipper (S2, EV) (Orchidaceae)

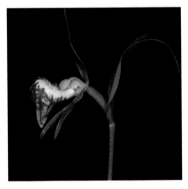

The rarest of the Adirondack lady's slippers, the ram's head lady's slipper has the smallest stature (less than one foot tall). Its flowers are also the smallest, but a close inspection reveals that they are the most extraordinary. As its common name suggests, the flower looks like the head of a charging ram. However, the common name does not convey the unique red-violet to red venation of its strongly angled pouch, which is about the size of one's thumbnail. The ram's head lady's slipper blooms in late May to early June, occurring in wet to dry conifer forests, often associated with calcium-enriched soils and bedrock, and tending to bloom more regularly near the edge of evergreen forests. Non-flowering individuals, comprising much of the population in deeply shaded habitats, look like smaller versions of the more common yellow lady's slipper (*C. parviflorum*), often in the same habitat and blooming at the same time. The leaves of the ram's head are typically smaller, narrower, and more bluish green than those of the yellow lady's slipper. If you find this species, look more carefully for the dozens of additional individuals likely to occur at a site, sometimes many feet away from each other.

Arietinum means "like a ram," as in the zodiacal Aries. Look for this species in forests with abundant northern white-cedar (*Thuja occidentalis*) and white spruce (*Picea glauca*).

Cypripedium reginae Walt., showy lady's slipper (EV) (Orchidaceae)

Reginae means "of the queen" —no other species name could be more appropriate for this lady's slipper species, the largest of all eastern US species of this genus and widely regarded as one of the most beautiful of all wildflowers native to the United States. The common name of just "showy" does not fully do justice to this species when in flower in early summer, especially in the large numbers occurring under the best conditions. This plant is so extraordinary in flower that it was historically collected intensively for the florist trade in places like northwestern Indiana. Even enamored botanists have taken their toll on this species with their past collecting practices.

The showy lady's slipper occurs in forested swamps, wet meadows, and wet roadside ditches as long as the water is enriched with calcium—conditions usually found in areas that have limestone bedrock or glacial till. (It is the surpassing beauty of this and other wild orchid species that explains why so many botanists and ecologists have been "driven into the ditch.") Individual plants can reach three feet in height. Its most robust flowering occurs under the brightest conditions, blooming halts entirely under dense forest canopies that lack gaps. The streaked coloration of the flower pouch is quite variable, ranging from the lightest pink (even pure white) to cherry or purplish-red. The yellow lady's slipper is also rather common in these habitats but blooms about three weeks earlier. Whereas the other lady's slipper species are pollinated by bees, the showy lady's slipper is pollinated by flies and beetles. The showy lady's slipper is the state flower of Minnesota and provincial flower of Prince Edward Island. There is no better reason to explore cedar swamps in early summer—not an easy trek while battling black flies—than to find this plant in bloom.

Nonflowering plants resemble the false hellebore (*Veratrum viride*), which often grows nearby but lacks the dense pubescence along the stem. Of all the lady's slipper species in the Adirondacks, this one is the best known for causing dermatitis among people who handle its very hairy leaves and stems.

The showy and yellow lady's slippers are particularly susceptible to deer browsing because of their stature and leaves
along the stems. Cedar swamps, excellent habitats for both species, are also prime habitats for white-tailed deer, which have
decimated many once-large populations of these lady's slipper
species in New York. Beaver also occasionally, but indirectly,
threaten these species when they build dams that flood these
swamps, killing not only the northern white-cedar but the
many different plants characteristic of these forested wetlands.
So while these species are legally protected from people, they
are not protected from other sources of elimination.

Neottia cordata (L.) R. Br. ex Ait. f., heart-leaved twayblade (EV) (Orchidaceae)

Occurring in forested wetlands
typically dominated by mosses
and conifers, especially black
spruce (*Picea mariana*) and
tamarack (*Larix laricina*), the
heart-leaved twayblade is the
most common of the *Neottia*
species in the Adirondacks.
It can also occur in northern
white-cedar (*Thuja occidentalis*) swamps where groundwater
is influenced by calcium. Even at sites where it occurs, typically as scattered individuals, it is still not easily seen because of
its diminutive stature, its pair of small, heart-shaped (cordate)

leaves, and its little dark red or purple to green flowers, which open in late spring into early summer. Because of the flowers' small size and foul fragrance, fungus gnats are the primary pollinators. This species also occurs in northern Asia, Europe, and Greenland, and at high elevations in western North America.

Pogonia ophioglossoides (L.) Juss., rose pogonia (EV) (Orchidaceae)

Rose pogonia is common in open, acidic peatlands (bogs) and saturated, acidic sands in meadows and ditches; in rich fens, it is usually sited on acidic hummocks, removed from direct groundwater influence. Its common associate is grass pink (*Calopogon tuberosus*), and both species can be quite abundant (over one thousand) at a location, flowering together in early to midsummer. The most exciting places to see these orchids are on quaking mats of bogs, a treacherous walking experience similar to walking on a water bed, except for occasional opportunities to fall through the mat into the rather deep water below. But these species are also common on much firmer peat substrates. Their local abundance is likely due to its nature of reproducing by vegetative means, sprouting new plants from its roots, as well as by seeds. We have estimated that there are hundreds of thousands of plants of both species spread across acres of saturated iron mine tailings in the western Adirondacks (see Bramen 2017).

Rose pogonia's single flower per stalk (sometimes two), colored from faint to dark pink (rarely white), is fragrant, but descriptions of its scent vary greatly, from pleasant to (in Henry David Thoreau's opinion) offensive (cited in Chapman 1997). Developing shrub and tree cover eliminates both rose pogonia and grass pink from bogs. *Pogonia* means "beard," referring to the appearance of the lip.

Epifagus virginiana (L.) W.P.C. Barton, beechdrops (Orobanchaceae)

"See those beech trees," said the instructor. "Those things under it are sons-of-beeches." This helps students to remember the easily overlooked parasite, *Epifagus virginiana*, commonly called beechdrops. As the genus name indicates, it grows on the roots of beech, *Fagus grandifolia*.

Beechdrops do not emerge in the Adirondacks until midsummer, when the stiff, narrow, dark-purple stems are barely discernible in the leaf litter. It lacks chlorophyll and is dependent on its host for water and nutrition. Or is it? Hundreds, if not thousands, of research hours have been expended trying to germinate seeds. All to no avail. Perhaps this is because beechdrops are somehow involved with the mycorrhizal fungi in the fine roots of the beech.

But it is not only the host-parasite relationship that is fascinating. Beechdrops have a very distinct floral syndrome. Most flower stems contain a mixture of both chasmogamous and cleistogamous flowers. Chasmogamous flowers open up (think "chasm") and can be visited by insects. Cleistogamous flowers (think "closed"), on the other hand, do not open but still produce abundant seeds without cross-pollination.

The seeds are dustlike. Raindrops play an important role in their distribution after the capsule opens, splashing thousands of seeds to their new homes.

Euphrasia stricta D. Wolff ex J. F. Lehm.,
upright eyebright (Orobanchaceae)

There is a lot of taxonomic confusion over the correct name for this widespread Adirondack plant of lawns, pathways, and disturbed sites. True to its extensive northern distribution, it likes cooler sites. Plants are annuals, seldom more than a few inches tall. Leaves are opposite, more or less round or oval, and have prominent teeth. Flowers are light purple or lavender and three-lobed. Easily overlooked, the flowers are attractive; they have a yellow throat. An American

Olympic swimming gold medal winner was apparently named after this plant—Euphrasia Louise Donnelly (1905–63).

Superficially, eyebright resembles a species of speedwell (plants in the genus *Veronica*). Among several differences, however, between the two groups of plants is the parasitic behavior of eyebright. Although it is green and can produce most of its own food, it forms parasitic attachments to neighboring plants, siphoning off food and nutrients via modified roots called "haustoria." Eyebright's haustoria are tiny but can be seen by examining neighboring roots of host species with a hand lens.

Various species of eyebright are widely sought after for use in homeopathic medicine for a wide range of remedies.

Capnoides sempervirens (L.) Borkh., pink corydalis (Papaveraceae)

Traditionally included in the genus *Corydalis* as *C. sempervirens* (L.) Pers., this northern inhabitant of rocky slopes, talus, and open woods is distinguished by its yellow-tipped flowers. These appear in late spring or early summer. In some parts of the eastern United States, pink corydalis is abundant after fires in forests. Leaves are compound with at least three orders of division of the rounded lobes or leaflets and are arranged alternately. The entire plant is often strongly glaucous (dull gray green in color, because of a layer of wax).

Polygaloides paucifolia (Willd.) J. R. Abbott, gaywings (Polygalaceae)

Often mistaken for an orchid, gaywings is a common perennial in dry to moderately moist forests, which usually have a sparse ground cover. It has a thin rhizome that gives rise to stems no more than six inches tall, with leaves clustered near the tip. The purple flowers are found at the top of the stem. They look like a miniature bird with wings and a feathery head. Bumblebees are the chief pollinators. At the base of the stem, and more commonly underground, are inconspicuous cleistogamous flowers that lack all the charisma of those on the stem. Cleistogamous flowers are devoid of showy structures, self-pollinate, and provide asexual reproduction.

Recent botanical research has moved this species from the genus *Polygala* to *Polygaloides*. *Polygala* means "much milk," an allusion to the idea that some species in this genus increase milk production in cows.

Persicaria amphibia (L.) Delarbre ssp. *laevimarginata* (Hult.) Soják, American water smartweed (Polygonaceae)

The most colorful of the numerous smartweeds that occur in the Adirondacks, American water smartweed is a true aquatic. It can grow in standing water and produce floating leaves but it also survives in wet soil, where it grows as an emergent. In wastewater ponds, this species was shown to accumulate a higher concentration of arsenic than other aquatic plants.

Leaves are lance-shaped, alternate, and at each place where a leaf joins the stem is a papery sheath, an ocrea, that distinguishes almost all plants in the smartweed family, the Polygonaceae. They are called smartweeds because of the pungent taste of most of the leaves. In mid- to late summer, it bears pink or red flowers crowded on a dense spike. This smartweed favors full sun in open wetlands, usually in standing water of ditches and slow-moving streams. American water smart-weed occurs in north temperate regions of both the New and Old World.

Persicaria maculosa Gray, lady's thumb (Polygonaceae)

Similar to the previous species, except leaves have a dark-purplish blotch in the middle of the leaf.

Persicaria pensylvanica (L.) M. Gomez, Pennsylvania
smartweed (Polygonaceae)

Similar to the previous species, except that it is common in weedy
areas and grows up to four feet (rather than three feet) tall.

Purple to Blue Flowers

Cichorium intybus L., chicory (Asteraceae)

 Like its relative dandelion,
chicory is a plant originally in-
troduced as food. Unlike dandelion, chicory culture has expanded, and chicory inulin is
now an important component
of several processed foods. The

roasted and ground roots have long been used as a coffee substitute or coffee additive. The young leaves are sold as a salad
green, and the trendy salad ingredient radicchio is derived
from a variety of chicory.

But in the Adirondacks, chicory is best known as an attractive roadside weed, one of the few with sky-blue flowers.
Chicory is a perennial, up to four feet tall, with alternate, dissected leaves and a rough-to-the-touch stem. The flowers are
actually clusters of flowers, between twelve and twenty in one
flower head. Flowers open in full sun and generally close by mid-
afternoon. The fruits are single-seeded and dry.

Eurybia macrophylla (L.) Cass.,
large-leaved aster (Asteraceae)

The large-leaved aster is common in moderately moist, open
woods and mature forests. Flower heads are about one inch in

diameter and consist of eight to twenty violet to pale blue (rarely white) ray flowers and many yellow disk flowers in dense, flat-topped clusters in late summer into fall. It is one of the first asters to bloom.

It is known by some as "lumberjack toilet paper" because of its large (*macro*, meaning "large," and *phyllos*, meaning "leaf"), broad, basal leaves (up to eight or more inches long), which are heart-shaped, pointed, and long-stalked. Basal leaves mostly disappear by flowering time. Plants can reach four feet in height and typically form dense patches from rhizomes.

Symphyotrichum cordifolium (L.) G. L. Nesom, heart-leaved aster (Asteraceae)

The heart-leaved aster occurs along forest edges and trails and in open woodlands. Flower heads are about five-eighths of an inch wide and consist of eight to twenty pale blue to purple ray flowers that surround the central yellowish to reddish disk flowers. Heads are in a rounded cluster. Flowering occurs from late summer into fall. Leaves are alternate along the smooth stem, heart-shaped, sharply toothed, hairy below, and stalked; basal leaves can be six inches long. Plants can reach four feet in height.

Echium vulgare L., viper's bugloss (Boraginaceae)

A field of this plant in flower is a beautiful sight. The masses of bright blue flowers immediately catch the eye. *Echium vulgare* is widespread in North America and grows in dry soils of fields, pastures, roadsides, railroad rights-of-way, and other disturbed and waste areas.

Viper's bugloss, so called because it was thought to cure snakebite, is native to the Mediterranean region. A biennial that usually forms a rosette of leaves the first year and then a flowering stem the second, it is a densely hairy plant. This feature is the source of the name "bugloss," a corruption of the Latin word for an ox's tongue, which has a rough surface. The hairs are quite stiff (hispid) and can puncture skin. The flowers are slightly two-lipped, with the upper part larger than the lower. Flowers are produced in a distinct pattern, an uncoiling that resembles a fiddlehead. As the flowers age, they lose much of their color and become pink.

Honeybees are among the pollinators, and viper's bugloss is an important nectar source. However, like numerous other plants, this species produces pyrrolizidine alkaloids that are toxic and have been reported to induce liver tumors. This toxin is present in the nectar and the resultant honey.

Brasenia schreberi J. F. Gmelin,
watershield (Cabombaceae)

This distinctive aquatic plant with floating oval-shaped leaves is one of the most common aquatic plants in the Adirondacks and many other parts of the world. It inhabits lakes, ponds, and quiet margins of streams in water six feet or less in depth. Watershield is the only common aquatic plant with peltate leaves, that is, leaves with the leaf stalk attached in the center of the leaf, thus giving the appearance of a shield and, hence, the common name. Most populations have leaves riddled with tunnels formed by leaf miners, often dug by the water lily leaf beetle (*Galerucella nymphaeae* L.). The plant seems to thrive in spite of its voracious guests.

The sex life of watershield is literally up and down. The flowering cycle usually spans three days. On day one, the flower bud emerges, then submerges at night. On day two, the six purple petals open, and the flower is in the female (pistillate) stage, with pollen in the anthers not released, thus ensuring that any pollen that does arrive is from a different plant. Then, the flower submerges for the night to emerge the next day in its male form. At this stage, the pollen is released from the stamens and carried by air currents to other flowers. At the end of day three, the flower again submerges for good, and the fruits develop underwater. Fruits are small, somewhat larger than a grain of rice, and usually contain two seeds. As the stalk decays, the fruits are dispersed by water currents.

The undersurface of the mature leaves and all the developing parts of the plant are covered with a thick, clear slime whose function is unclear. Unpalatable as they may look and feel, these slime-covered parts of the plant are considered a delicacy in Asia, where this perennial aquatic is grown as a vegetable.

Campanula rotundifolia L.,
harebell (EV) (Campanulaceae)

Like several other plants that find their home in the Adirondacks, harebell is one of the guild of species of northern and arctic regions around the globe. It is a delicate, winsome plant of rocky outcrops, alpine meadows, and cliff faces.

It is given the name *Campanula* because the blue flowers are bell-shaped. The five-lobed corolla is visited by bees. The specific epithet means "round leaves," in reference to the basal leaves that are round in outline. These usually wither rapidly as the season progresses, leaving only the hairlike stem leaves visible. The harebell is state-protected because it is considered exploitably vulnerable.

Lobelia dortmanna L., water lobelia (Campanulaceae)

Water lobelia often indicates good water quality, so look for this delicate aquatic plant in clean, clear lakes and ponds, especially those with a sandy bottom. Like several other submersed aquatic plants with a rosette of basal leaves, water lobelia functionally is referred to as an isoetid, because of its similarity to quillworts (genus *Isoetes*), also abundant in Adirondack waters. Isoetids have a distinct form of photosynthesis and are able to absorb carbon directly through their roots.

Seeds of this perennial germinate under water up to six feet deep. Deeper water hinders the quality of light needed for germination. Each plant produces a single flowering stalk that has up to ten light-purple to white flowers about an inch long. Some flowers develop and produce seeds under water. Seed production is prodigious in both emersed and submersed flowers. Unlike the majority of aquatic plants, water lobelia has no form of asexual reproduction.

Lobelia inflata L., Indian tobacco (Campanulaceae)

Native Americans used the plant to treat asthma. It contains lobeline, a compound that mimics the effects of nicotine. Research has shown that lobeline may actually reduce the effects of nicotine in the body. While ingesting large quantities of the plant is toxic, the small quantities used in recommended herbal remedies are considered safe for most people.

Indian tobacco is a hairy annual growing to a height of two and a half feet, with alternate, oval-shaped leaves. Attractive blue to white flowers are produced at the top of the stem. These have five lobes—two upper, three lower. It grows in open sites, along paths, in meadows, and in other places with exposed soil. A distinctive feature of the plant is the swollen calyx that develops as the plant matures. None of the other six species of Adirondack lobelias have this calyx feature.

Lobelia kalmii L., Kalm's lobelia (Campanulaceae)

This perennial grows to about sixteen inches and is our most delicate lobelia, with weak stems and attractive pastel blue flowers that have a white center. The two upper lobes of the corolla are considerably smaller than the lower lobes. The stem leaves are linear and different from the larger basal leaves, which are usually withered by the time the plant flowers. Its habitat is specialized, favoring rich fens (open wetlands that have calcium-enriched groundwater).

The species is named after Pehr Kalm (1715–79) a botanical explorer of North America and a student of Carl Linnaeus. Kalm visited the Adirondacks in the mid-1700s.

Lobelia siphilitica L., great blue lobelia (Campanulaceae)

This is the largest of the blue-flowered lobelias in the Adirondacks. It is a tall—up to four feet—stately perennial with crowded alternate leaves. Larger, more basal leaves have hardened white teeth (mucronate teeth). Flowers are an inch across and, like all lobelias, have two small upper lobes and three larger lower lobes. They appear in the late summer and fall. This is a wetland plant found in forest seeps, low road-sides, and streambanks. Its bright blue flowers are a colorful contrast to the yellows and oranges of so many fall wildflowers.

Great blue lobelia contains the alkaloid lobeline, found in all members of the genus. This may be why it was thought to be a cure for syphilis, as suggested by the specific epithet.

Lobelia spicata Lam.,
common spiked lobelia (Campanulaceae)

A perennial of moist to wet sites in old fields, along road-sides, and in other disturbed areas, this native plant is also found along rivers and in rocky outcrops. Most leaves are basal, egg-shaped (broadest at the tip), and hairy. Those on the stem are smaller in size. The plant grows to over three feet tall.

In midsummer a dense spike (hence the specific epithet *spicata*) of pale blue flowers is produced. A dozen or more flowers may be open at one time. Like other lobelias, there are two small upper lobes (these usually recurved) and three larger lower lobes on the corolla. When the buds begin to open, these flowers are actually upside down. In an interesting twist during development, the flowers turn 180 degrees around.

The style in this species is blue. As noted for cardinal flower, style behavior in *Lobelia* species is highly specialized. The anthers form a tube around the style, and as the style expands, it pushes the pollen outward, where it can be picked up by a pollinator. It is only after this, when it becomes certain that the style must receive pollen from flowers other than itself, that the style becomes receptive.

Gentiana linearis Froel., narrow-leaved gentian (Gentianaceae)

The most intensely blue of any Adirondack wildflowers, narrow-leaved gentian is common in marshes and stream and lake borders, where its brilliant blue contrasts with the surrounding vegetation. It is an herbaceous perennial with opposite leaves (like all gentians), growing to three feet tall. This species is one of the bottle gentians, plants that produce flowers that look as if they are in bud even when mature. Bottle gentian flowers require the strength of a large bumblebee to spread the lobes and pollinate. A capsule of many small seeds is produced in the late summer.

Chewing on one of the leaves makes it clear why species of this genus are an ingredient of the bitters used in mixed drinks.

While all parts of the gentian plant are strongly bitter, commercial production of bitters is derived from its roots.

Iris versicolor L., blue flag (Iridaceae)

Iris pseudacorus

Frequent in a variety of wetland situations throughout the Adirondacks, blue flag is the most widespread native iris of the region. It grows throughout the eastern United States and eastern Canada.

Irises have a unique morphology. The thickened rhizomes are often on or near the soil surface. Leaves arise from the rhizome, overlapping in a single plane and growing up to three feet tall. The flowers also have a distinct structure. What might at first appear to be three separate flowers is actually a single flower with three separate pollination units. This allows pollen from different pollination units to pollinate a different unit in the same flower. Blue flag flowers are among the largest in the Adirondack flora, up to five inches across and blue or violet. It flowers from midspring to summer.

Native Americans used blue flag for a variety of medicines. However, caution is advised when ingesting the plant because of the presence of iridins, known to cause poisoning in humans. Likewise, gloves are needed when handling leaves and rhizomes because of the prevalence of abundant crystals (raphides). These raphides are needle-shaped and can easily penetrate skin, causing not only pain from the puncture and oxalates but also from toxins contained in the plant. In experiments to control runoff from golf courses, where large quantities of pesticides and herbicides are used, blue flag was able to sequester more of the toxins than other plants.

In many shallow waters throughout the Adirondacks, especially near towns and villages, the naturalized yellow iris, *Iris pseudacorus*, can be abundant. This species is more robust than blue flag, and when its yellow flowers open, it cannot be mistaken for any other native or exotic species.

Sisyrinchium angustifolium Mill., narrow-leaved blue-eyed grass (Iridaceae)

Flowers of species of this genus do not bear a recognizable resemblance to irises even though they are closely related. But the arrangement of the leaves in the two groups of plants is similar: both have flat leaves that overlap in the same plane. This "equitant" condition is most prominent and best known in the garden iris.

Narrow-leaved blue-eyed grass flowers in early spring in open, sunny areas in dry woodlands, fields, and meadows. This species has narrow, grasslike leaves, about a foot tall. Flowers are dark blue to blue with a yellow center and produced on several stalks. Each petal has an indentation at the tip with a short, hairlike projection in the middle of the indentation. The fruit is a round capsule with small seeds. This is our largest blue-eyed grass and widespread across much of the United States and Canada.

Sisyrinchium montanum Greene,
mountain blue-eyed grass (Iridaceae)

Usually no more than a foot tall, this blue-eyed grass has a single flowering stem with light-blue to blue flowers that have a yellow-green throat. Each of the flower segments has a needle-like tip. Leaves are grasslike in appearance and arranged like those of an iris, which is in the same family. Mountain blue-eyed grass favors damp, open, sunny soil in meadows and along shores of lakes and streams. It is frequent in the Adirondacks and occurs throughout much of eastern North America and Canada.

Mentha canadensis L., American wild mint (Lamiaceae)

Wild mint favors wet areas such as marshes, shorelines, streambanks, and wet meadows. It grows to about two feet tall, with flowers in whorls encircling the upper stem. Flowers are lilac (rarely white), two-lipped, and quite small. Leaves are elongate and toothed. On a hot day, it is often possible to smell the plant before seeing it. All parts of the plants are edible and can be used in the same manner as fresh mint leaves purchased in stores. Perhaps the most widespread use of the plant is a preparation of tea from the dried leaves.

By some it is considered synonymous with *Mentha arvensis* var. *canadensis*.

Prunella vulgaris L., self-heal (Lamiaceae)

Ubiquitous in lawns, road-sides, fields, paths, and other disturbed areas, self-heal is taxonomically confused. Some botanists recognize two species; others, a single, wide-ranging variable species. There is also disagreement over whether it is part of the native flora or an introduction. Like all members of the mint family, self-heal has opposite leaves and a square stem. It survives mowing, so it is frequent in parks and on roadsides. In these abusive situations, plants are often stunted, and flowering stems are decapitated. Flowers are produced in midsummer. They are two-lipped; the upper one is purple, and the lower one is white.

All parts of the plant are edible but are best collected in the early spring (and from a site that is not doused with herbicide). Naturopaths have long favored this plant for a variety of uses. As one medicinal plant website states, "This is a plant that can help many health ailments."

Scutellaria galericulata L., marsh skullcap (Lamiaceae)

Scutellaria lateriflora

The common name "skullcap" refers to the prominent bump on the calyx of the flower. All species of the genus have this structure, which bears a fanciful resemblance to the caps worn by monks. The genus name references the word "dish," perhaps a more apropos analogy but not used in any common name. The specific epithet *galericulata* means "hooded," alluding to the shape of the flower. So this is the dish cap plant.

Marsh skullcap is widespread in northern regions around the globe, occurring in marshes and along lakeshores in thickets of sweet gale (*Myrica gale*) and speckled alder (*Alnus incana* ssp. *rugosa*) as well as wet meadows. Stems are weak and often lax. Leaves are opposite, arising from a square stem that often roots at the base. Flowers are blue, small (one inch), covered with soft hairs, and produced in the axils of leaves. They often are found all on one side of the stem. Like all members of the genus, it is not aromatic even though it is a member of a family renowned for its aromatic constituents.

Another skullcap in the Adirondacks is mad-dog skullcap, *Scutellaria lateriflora* L. It grows in habitats similar to those that are home to marsh skullcap, though it is less shade tolerant and is also distributed globally in northern regions. The two skullcaps can be distinguished from one another by the location of the flowers. Those in marsh skullcap are in the axils of the leaves, and the leaves lack leaf stalks. Flowers of the other species are borne on lateral branches.

Utricularia purpurea Walt.,
purple bladderwort (Lentibulariaceae)

The diverse waters of the Adirondack lakes, streams, and marshes provide habitat for about ten species of one of the most fascinating groups of plants in the flora, the bladderworts. All species are obligate aquatic plants, meaning that they require standing water or saturated soil. The bladder is the distinguishing feature. It is a small saclike structure with a trapdoor that, when stimulated by a small animal, suddenly opens and captures its prey.

Purple bladderwort is one of the truly aquatic species, occurring unrooted in still water and growing just below the surface. Branches occur in whorls on the stems and bear capillary segments with bladders on some of the tips. The only time the plant is evident is in midsummer, when the purple flowers extend out of the water. Bladderworts are innovative regarding nutrition, capturing animals to supplement their nitrogen, but when it comes to reproduction, they are old-fashioned and depend on insects for pollination. However, there is some evidence that the bladder's main function is not trapping food but providing an incubator for microorganisms that help feed the bladderwort.

Purple bladderwort is distinctive among Adirondack bladderworts in having whorled branches with bladders at the tips of some branches. Reclined bladderwort, *Utricularia resupinata* B. D. Greene ex Bigelow, also has purple flowers but is rooted and much smaller. It lacks above-ground branches.

Trillium erectum L.,
purple trillium (EV) (Melanthiaceae)

The purple trillium occurs in rich, moderately moist, upland forests of hardwoods and occasionally scattered eastern hemlock (*Tsuga canadensis*) and eastern white pine (*Pinus strobus*). It is also known as "wake-robin," because its flowers greet robins as they migrate north in the early spring, and "stinking Benjamin," because of its flowers' foul scent that attracts carrion flies that pollinate this species. The color of its flower, with its velvety texture, can be quite variable, from reddish-purple to dark red, purple, yellow-green, and even white. Its flowers can be erect or drooping, but if the latter, still typically held above or at the level of the leaves. The ovary of purple trillium is maroon, and its fruit is a slightly pyramidal, dark maroon berry.

Epipactis helleborine (L.) Crantz, helleborine (Orchidaceae)

The helleborine is the only non-native terrestrial orchid in New York and the only species of orchid not legally protected in the state. This species was introduced from Eurasia; discovered in North America in woods near Syracuse, New York, in 1879; and is now widely naturalized in the eastern United States. It is also known as the "weed orchid," and is the least habitat-specific and most common of all orchids in New York, occurring in lawns, garden beds, old fields, roadsides, and forested uplands and lowlands.

The helleborine blooms during a good portion of the summer, with up to fifty flowers (usually fifteen to thirty-five) per stalk. Individual flowers look similar to the common epiphytic and tropical *Cymbidium* orchids sold around Easter but are significantly smaller. Plants are quite variable in height, typically from twelve to thirty-six inches tall. Shorter-stature, nonflowering plants are easily confused with *Cypripedium arietinum* and *C. parviflorum*. Throughout Eurasia, the rhizomes and roots of this species have been used for centuries to treat medical conditions, including gout, which might explain its introduction into the United States.

Galearis spectabilis (L.) Raf., showy orchid (EV) (Orchidaceae)

The showy orchid blooms in small to large patches in midspring in rich, moist upland, deciduous forests in soils that are often enriched by calcium. It thrives in ravines where moisture conditions are ideal and in forests that have canopy openings. Sugar maple (*Acer saccharum*) is usually the most dominant tree, while white trillium (*T. grandiflorum*), hepatica (*Hepatica nobilis*), and many other wildflowers are often associated with this orchid in the ground layer.

The species name *spectabilis*, means spectacular or showy, referring to the flowers. Flowering plants have two basal leaves, usually intermingled with younger, single-leaf, nonflowering individuals in close groups.

Platanthera grandiflora
(Bigelow) Lindl., large
purple fringed orchid
(EV) (Orchidaceae)

Platanthera psycodes
(L.) Lindl., lesser purple
fringed orchid (EV)
(Orchidaceae)

The two purple fringed orchids of the Adirondacks occur in lowland forests, along moist, disturbed areas near forest edges and streams, and in wet meadows and ditches. The differences between these species are subtle but real. The large purple fringed orchid is more shade tolerant; larger (reaches over three feet tall); and has fewer, less dense but larger, and deeper purple flowers that bloom earlier (early to midsummer) than those of the lesser purple fringed orchid (blooms mid- to late summer). The single most reliable way to separate these species is by the shape of the opening to the spur; it is circular in *P. grandiflora* and shaped like a transverse dumbbell in

P. psycodes. These seemingly trivial differences result in different insects pollinating each species.

The species names refer to the shapes of the flowers: *grandiflora* is large-flowered; *psycodes* is butterflylike. Contrary to their common names, flower color can vary quite a bit—from white to lilac to purple. They can occur together.

Orobanche uniflora L., one-flowered broomrape (Orobanchaceae)

This fascinating plant is probably more frequent than reported because of its furtive habit of often flowering under leaf litter. It is usually described as being an annual, but the term "monocarpic" (fruiting only once) is more accurate, since the plant can survive for several years underground before flowering. No chlorophyll is present because the plant is totally dependent on its host for food and water. Since it is parasitic, its only reason for emerging is to flower.

As the specific epithet indicates, there is a single flower, arising from a white, swollen, tuberlike structure on a hairy flowering stalk that is up to five inches tall. Flowers are five-parted and pale blue to white in color, with two yellow pollinator guidelines on the lower lip. Flowers appear in early spring and are soon followed by a capsule that is often covered by the dried, persistent flower. Seeds are dustlike and will only germinate in the presence of chemicals produced by suitable hosts. This is a sophisticated communication system, with both the host and the seed sending and receiving signals.

Hosts are from diverse families, though plants from the aster family appear to be favored. The genus name *Orobanche* means "bean strangler," chosen because some Mediterranean species are serious parasites of legumes. The odd common name "broomrape" refers to well-known European species that parasitize broom [*Cytinus scoparius* (L.) Link].

Mimulus ringens L., Allegheny monkey flower (Phrymaceae)

Mimulus reaches its greatest diversity in the western United States, with only one wide-ranging species frequent in the Adirondacks. Because the corolla has a fanciful resemblance to a grin, the genus was named after the Latin word for a clown or mime.

A perennial that grows to a height of four feet, monkey flower has a distinctly square stem and opposite leaves. The purple flowers are on long stalks that arise from the axils of the leaves. This is a plant of open wetlands, ditches, marshes, and gravel and sand bars of streams, and it is widespread in most of the United States. It flowers from midsummer until frost.

A tea made from the dried leaves is purported to be a general tonic.

Pontederia cordata L., pickerel weed (Pontederiaceae)

Ubiquitous in Adirondack lakes, streams, and marshes, pickerel weed is a true aquatic that requires standing water to thrive. This is reflected in its common name, taken from the fish that is often found where this plant is abundant. The bright blue masses of flowers produced throughout the growing season are some of the most attractive of our aquatic plants. It is named for Italian botanist Guilio Pontedera (1688–1756).

The heart-shaped (cordate) leaves are similar to those of several other aquatic plants but can be distinguished by uniformly sized veins that all start at the base of the leaf and curve up to its tip. Each flowering stalk arising from the thick underground rhizome is subtended by a single leaf. Leaf morphology can be strongly influenced by water conditions, so leaves can be either heart-shaped or strap-like. Flowers are two-lipped. The upper lip is three-parted, with a bright yellow dot in the center lobe. The lower portion of the flower is also three-lipped but with narrower segments.

As the fruits begin to mature, the flowering stalk bends so that the single-seeded fruit develops under water.

Houstonia caerulea L., common bluets (Rubiaceae)

Common bluets are also known as "Quaker ladies" because of their modest appearance or perhaps because the four-parted flower resembles the hats worn by Quaker ladies of an earlier era. In any event, this delightful spring flowering perennial cheers roadsides, dry to moderately moist forests, and moss-covered rock. It only grows two or three inches tall.

Flowers are blue with a yellow center, ranging from a deep blue to almost white. Superficially, all the flowers appear similar but a closer look reveals that some have a slender style emerging from the orifice of the corolla while other plants lack the obvious style and have protruding stamens instead. This production of two different kinds of flowers in the same species is called "heterostyly" (different styles). Heterostyly ensures cross-pollination from short-styled flowers to long-styled flowers and in the opposite direction as well. This widespread phenomenon in plants is immortalized in Darwin's very readable book *The Different Forms of Flowers on Plants of the Same Species*, first published in 1877 and still in print.

Verbena hastata L., blue vervain (Verbenaceae)

This is a plant of moist or wet areas, found in marshes, wet fields, and especially on sand and gravel bars of streams. It is a tall (up to five feet), rigorous, and attractive perennial with opposite leaves that are usually double serrate, that is, the larger teeth have smaller teeth between them. The entire plant is covered with hairs that are rough to the touch. In late summer or fall, the blue (rarely white) flowers appear on elongated, pencil-like spikes. Each of the small flowers is slightly two-lipped, and the five flower lobes are all approximately the same size and shape.

An archaic name for blue vervain is "simpler's joy," where "simpler" does not mean uncomplicated or lacking in wisdom but rather refers to a medicine, or "simple." The leaves of the plant are used for tea, purported to address several ailments.

Viola cucullata Aiton, marsh blue violet (Violaceae)

A true wetland plant, this is a stemless blue violet found in saturated hollows and seeps in bogs, fens, and swamps. Leaves are heart-shaped. The hairy cleistogamous (inconspicuous, self-pollinating) flowers are produced on stalks almost as tall as the leaves.

Viola sororia Willd., common blue violet (Violaceae)

Viola adunca   *Viola rostrata*

This is indeed a common violet, found throughout the eastern half of the United States and Canada. *Sororia* sounds like "sorority" and comes from the same root, "sisterly," applied to this species because it resembles several "sister" species. The cleistogamous (hidden) flowers in this species have an elastic capsule that ruptures and forcibly propels the seeds away from the mother plant. Common blue violet has distinctly heart-shaped leaves and may be completely hairless or densely hairy. It is found in a variety of sites, including forests, lawns, pastures, and other disturbed sites, and is thus easily grown as a garden subject.

Like most violets, the young leaves are edible, though bland. The flowers of this species in particular, however, make delicious jelly.

Three additional stemmed violets with blue to purplish flowers that are similar in appearance occur in distinct habitats in the Adirondacks, especially relative to soil moisture: *Viola adunca* Sm., sand violet; *Viola labradorica* Schrank, American dog violet; and *Viola rostrata* Pursh, long-spurred violet. The sand violet often occurs on well-drained sandy sites or in dry woods and alpine ravines. The American dog violet occurs in moist to wet woods and meadows, and along streambanks. The long-spurred violet can be found among the spring wildflowers of rich, moderately moist hardwood forests. The spurs of long-spurred violet are the longest of these three species, and its petals are hairless inside.

White Flowers

Sagittaria graminea Michx.,
grass-leaved arrowhead (Alismataceae)

Widespread in eastern North America, grass-leaved arrowhead grows in marshes, margins of lakes and streams, and as a submergent in lakes and ponds. Well described by the common name, this aquatic perennial can have bladeless leaves (phyllopodes) that look like grass. Phyllopodes are more common when the plant grows submerged in standing water. When growing as an emergent, the leaves are about eight inches tall; they are usually much longer when the plant is a submergent. Emerged leaves have an expanded blade. Like other members of the genus, the plant has white latex.

Flowers are produced on a stem up to two feet high with whorls of flowers, each with three white petals. Upper flowers are male (staminate); lower flowers contain both male and female components, though in some plants all the flowers are staminate. Fruits are one-seeded and grow in a cluster. They are an important food for waterfowl.

Sagittaria latifolia Willd.,
common arrowhead (Alismataceae)

Arrowhead is widespread in diverse standing water and mud
habitats. The common name describes the usual shape of the
leaf—like a broad arrowhead. But leaf shape is exceptionally variable, and it is possible to find plants with leaves a foot
wide and yet other arrowhead plants with very narrow leaves.
Sometimes in running water the leaves lose their arrow shape

and are strap-shaped. Whatever the shape, the leaves can always be distinguished by the white latex observed when a leaf is torn. Flowers are borne from mid- to late summer in whorls of three on the flowering stem, with the male flowers above the female flowers. Flowers of both sexes have three white petals that fall a few days after opening. The female flowers produce clusters of single-seeded fruits with a short beak.

In the past, arrowhead was a valuable food source not only for wildlife but for many groups of Native Americans. In the fall, the plant produces swollen tips of rhizomes that are rich in starch. About the size and shape of a small hen's egg, this specialized structure, a turion, has a well-developed bud. As winter approaches, the connection with the rhizome disintegrates and the turion can be carried by water currents or waterfowl to establish a new population.

Allium tricoccum Aiton, wild leek (Amaryllidaceae)

The five- to ten-inch-long lancelike leaves and reddish stalk of wild leek, also known as "ramps," emerge in rich, moderately moist, upland forests of deciduous trees in early spring, when numerous other wildflowers begin to bloom. It is commonly found in thick, large patches. Leaves senesce as the tree canopy fully leafs out. The rounded clusters of white flowers open in mid- to late summer above a leafless stalk about six to twelve inches tall, followed by a cluster of shiny, jet-black seeds. The seed stalk can persist into the fall. The species name comes from the Greek *tri* and *kokkos* ("three seeds"), referring to the number of seeds held in each fruit capsule.

Leaves and bulbs are edible, with the distinct odor and flavor of onion. Plant parts can be blanched, sautéed, or baked in a mixture with other ingredients. When harvesting this plant, one should collect only a very small portion of the patch, and not every year, to ensure its continued existence. It is increasingly sought after by specialty restaurants. Juice from the crushed bulb was historically used to treat insect stings.

Angelica atropurpurea L., purple-stemmed angelica (Apiaceae)

This is one of the tallest herbaceous plants of open, wet meadows and streambanks, arising from a thick taproot and growing up to eight feet tall. The large stems are hollow and, like the rest of the plant, smell like celery. Lower leaves are large and doubly compound, with each leaflet subdivided into three more leaflets. The base of the leaf wraps around the stem. Stems are purple (*atropurpurea* means "dark purple") and striated. In midsummer, large round clusters of greenish-white flowers up to eight inches across appear.

The name of the genus is derived from the word "angel," in reference to a cherished cordial made from a related plant in Europe. The young flower clusters are edible and tasty both raw and cooked but must be collected in early spring. Nevertheless, it is best to completely avoid eating this plant, because similar looking plants growing in the same habitat are among the most toxic in the flora, especially bulb-bearing water hemlock (*Cicuta bulbifera*).

Cicuta bulbifera L.,
bulb-bearing water hemlock (Apiaceae)

This is one of the most toxic plants in the Adirondacks and has been described as "violently poisonous." It is related to the poison hemlock, *Conium maculatum* L., that killed Socrates. The Greek philosopher's poison source was Eurasian, whereas *C. bulbifera* is native.

It grows in open, sunny areas along streams and other bodies of water as well as in sedge meadows and marshes—sometimes in standing water. Typically, it occurs singly, and large populations are rare. The compound alternate leaves are finely dissected into narrow segments. Like other members of this family this plant has swollen nodes (points on the stem where leaves emerge) and a flat-topped cluster of flowers. Flowers are white and often do not produce seeds. In the upper axils of the leaves, bulbils are produced. These are tiny budlike structures that break from the parent plant to establish a new plant.

Cicuta maculata L., spotted water hemlock (Apiaceae)

The most poisonous native plant in the Adirondacks, this wide-ranging perennial relative of carrot and celery has been implicated in several deaths. The common name is descriptive. *Maculata* means "spotted," which describes the stem. And the plant inhabits wetlands, where it is usually found scattered and not in dense stands. Leaves are alternate and divided into an odd number of leaflets. Bases of the leaves clasp the stem. Flowers are white and produced in flattened heads. Stems grow to six feet tall, often have purple blotches (maculate), are hollow, and are often striped.

The term "hemlock" is a reminder of the toxic nature of all parts of this plant. It is considered to be as dangerous as poison hemlock, *Conium maculatum*, which is also present in the Adirondacks and sometimes confused with water hemlock. The two can be distinguished, however, by the generally more robust nature of poison hemlock, by the division of the leaves, and its weedy nature. Further, poison hemlock leaflets are subdivided, while those of water hemlock are not. There are frequent reports of cattle being poisoned after grazing on water hemlock.

The fact that the same name—"hemlock"—is applied to the common gymnosperm eastern hemlock (*Tsuga canadensis*) and the poisonous angiosperms generates a certain amount of confusion. One explanation is that the appearance of the leaves of the tree reminded early botanists of the foliage of the poison hemlock, *Conium maculatum*. There is no relationship between the two groups. In fact, the hemlock tree has a long history of edible uses, including flavoring root beer.

Daucus carota L., wild carrot (Apiaceae)

Better known as Queen Anne's lace, this biennial is one of the most widespread and readily recognized roadside weeds. It also grows in old fields, pastures, and other disturbed areas. The delicate, lacy heads of white flowers appear in midsummer. Careful examination of each piece of "lace"—each flowering head—reveals a single red flower in the center—likened to a drop of blood drawn when Queen Anne pricked herself while making lace. An apt image, but what is the function of this single flower? Darwin was puzzled by the single red flower (sometimes there are a few, but always in the center of the floral display) and concluded that it serves no purpose. More than a century later it remains a mystery. Some researchers claim that the red flower serves to invite certain pollinators, but the proverbial further research is needed.

   With alternate, finely dissected leaves and densely hairy stems, Queen Anne's lace grows up to five feet tall and flowers from late spring to early summer. As the fruits develop, the flower heads assume a basket-shaped form.

Queen Anne's lace has a mixed record on edibility. It is literally carrot gone wild, as the fragrance of the foliage and root attest. It also contains some of the same compounds as wild parsnip, *Pastinaca sativa*, a serious purveyor of dermatitis, but in much smaller quantities. Nevertheless, anyone who is highly susceptible to photodermatitis should avoid handling and ingesting the plant. More important, if you use Queen Anne's lace, *be certain not to confuse it with poison hemlock*! Poison hemlock is deadly poisonous and has a fragrance similar to that of wild carrot.

Osmorhiza claytonii (Michx.) C. B. Clarke, bland sweet cicely (Apiaceae)

This is a spring-flowering perennial found in rich, moderately moist forests as well as well-drained soils in floodplain forests. Plants are about two feet tall and have alternate, compound leaves. Each leaf is divided into three leaflets, and each leaflet, in turn, has three parts. Flowers are white and produced in scattered clusters with a—collectively—more or less flattened appearance. Some flowers are male (staminate).

An anise fragrance does *not* characterize this species (in contrast to aniseroot, *Osmorhiza longistylis*)—hence, the appellation "bland." Other features separating the two are the number of flowers per cluster (fewer in this species, giving a more open appearance to the flowering branch) and the fleshiness of the root (more fleshy in *O. longistylis*). Bland sweet cicely seeds (technically fruits) have been used as an anise substitute, though the roots have been described as "rank tasting."

The colorful botanist Constantine (he was born in Constantinople) Samuel Rafinesque (1783–1840) apparently named the genus in honor of George Washington (*Washingtonia*) but earlier had already established the genus *Osmorhiza*. This confusion resulted in the rejection of the name *Washingtonia* for this group. Rafinesque apparently botanized in the vicinity of Lake George around 1819.

Sanicula marilandica L., Maryland sanicle (Apiaceae)

The Maryland sanicle, commonly known as black snakeroot, occurs in rich, moderately moist forests, floodplain forests, and in swamps on elevated microtopography. Flowers are one-eighth of an inch in diameter, white, and in small, rounded heads in a sparsely branched cluster that arises above the leaves. Flowering occurs in early summer. The fruit is oval, dry, and with recurved bristles. Leaves are palmately compound, consisting of five leaflets.

The name is thought to come from the Latin *sanus* ("healthy") because of the purported healing benefits of the plant.

Sium suave Walt., hemlock water parsnip (Apiaceae)

This frequent wetland plant is actually edible, even though it has toxic kin. Because of possible confusion of this plant with poison hemlock and spotted water hemlock, water parsnip should not be eaten. The common name "hemlock water parsnip" is an allusion to its deadly relatives. Hemlock water parsnip is a true wetland plant that can grow in standing water of marshes and stream and lake margins. It reaches up to six feet tall and has alternate leaves with leaf bases sheathing the stem.

Leaves are pinnately compound (with leaflets opposite one another on the axis of the leaf) and quite variable. Those growing in the water are sometimes deeply dissected. Small white flowers are produced in midsummer in flat-topped masses.

Hemlock water parsnip is most likely to be confused with spotted water hemlock, *Cicuta maculata*. Distinguishing features between the two include the larger, coarser teeth of water parsnip; the lack of purple splotches on hemlock water parsnip; and the distinct fragrance of hemlock water parsnip. It is related to the bulb-bearing water hemlock but has much wider leaf segments and does not produce bulblets in the axils of the upper leaves as *Cicuta bulbifera* does.

Apocynum androsaemifolium L., spreading dogbane (Apocynaceae)

This is an attractive perennial arising from a sturdy rhizome that produces annual stems that grow up to three feet tall. Leaves are opposite, oval-shaped, and have a smooth margin. It blooms in midsummer with numerous flowers appearing at the same time, giving the plant some value as a garden subject. Flowers are fragrant and white, with distinct pink stripes. Fruits are distinct, narrow, long podlike structures that open on one side to release seeds that have silky hairs, an adaptation to wind pollination.

Apocynum cannabinum

It is called spreading dogbane because it often branches. The plant can be aggressive and become a problem in tidy gardens. Native to much of North America, it grows in a diversity of habitats and tolerates light shade.

Spreading dogbane is used in homeopathic medicine but caution is essential when ingesting any part of the plant, owing to the presence of cardiac glycosides, substances that interfere with heartbeat. For this reason, the plant is seldom grazed by animals and is toxic to dogs, hence the common name. One odd feature of some medicines derived from this plant is an alteration of olfaction so that everything smells like honey.

Indian hemp, *Apocynum cannabinum* L., is also widely distributed in the Adirondacks. It seldom branches and has flowers with spreading lobes that are smaller than the flowers of spreading dogbane. The specific epithet *cannabinum* references the hemp (or marijuana) plant, *Cannabis sativa*. Both species of dogbane have strong fibers in their stems that were used by Native Americans for cordage.

Calla palustris L., wild calla (Araceae)

Like several other Adirondack plants, wild calla has a global distribution in the North Temperate boreal zone, where it inhabits cold water bogs, margins of lakes, and ditches. Flowers are very small and borne on a thickened axis, the spathe, that grows about a foot tall. The uppermost flowers on the spathe are usually bisexual, while those lower down are mainly staminate (male). Surrounding the spathe is a white modified leaf, the spadix. Wild calla is one of the few plants reported to be pollinated by snails. In addition, flies have been noted as pollinators. Leaves arise from the extensive rhizome at the base of the plant; they are somewhat fleshy and heart-shaped. In late summer and fall the bright red fruits appear.

Like most members of this family, the arum family or Araceae, calcium oxalate crystals are found throughout the plant, including the fruit. Calcium oxalate damages mucous membranes and can exacerbate the development of kidney stones. However, drying and boiling will destroy the crystals. The rhizome has been used as an emergency food in Scandinavia.

Aralia hispida Vent., bristly sarsaparilla (Araliaceae)
Older plants of this denizen of rocky, exposed habitats and disturbed areas are woody, making it technically a subshrub. Like other Adirondack *Aralia* species, it is native and has its flowers in umbels (globe-shaped clusters). Flowers of bristly sarsaparilla are small and white, yielding a fruit that looks like a blueberry. Leaves are pinnately compound, with five to seven elongate leaflets. The term "bristly" (*hispid* in Latin) refers to the stiff hairs on the stem, which are capable of puncturing skin.

True sarsaparilla is derived from a woody vine native to the Caribbean and its name is applied to this and other unrelated plants because of a similarity in the taste of the roots. Bristly sarsaparilla is seldom used for that purpose today but is widely sought after as an herbal remedy.

Aralia nudicaulis L., wild sarsaparilla (Araliaceae)

Sarsaparilla is best known as a root beer flavoring, and as a result, a group of unrelated plants with a similar taste incorporate "sarsaparilla" as part of their common names. True sarsaparilla was discovered by Columbus in the West Indies in his quest for spices. It is derived chiefly from the root of a catbrier (*Smilax regelii* Killip. and Morton). Wild sarsaparilla, *Aralia nudicaulis*, on the other hand, is an unrelated plant characteristic of the northern hardwood forests of the northeastern United States and adjacent Canada. In fact, it is considered an indicator species of some forest types in this region.

Also known as false sarsaparilla, this is a perennial with a strong, extensive rootstock from which the characteristic leaf arises. There are three leaflets on a stalk two feet tall. A shorter separate, leafless flowering stalk arises from the rootstock, giving the species its name, *nudicaulis*, meaning "naked stem." This stalk terminates in a globose mass of small, greenish-white flowers produced in the spring. In July, the fleshy black fruits ripen. Within a few weeks of fruit maturation, the leaf turns yellow and drops.

Prepared concoctions from this plant were widely marketed as a general tonic.

Hydrocotyle americana L., American marsh pennywort (Araliaceae)

This frequent occupant of wetlands might not bear much resemblance to English ivy, *Hedera helix* L., but recent studies have shown it is best placed in that family, the Araliaceae, which also includes ginseng among its members.

A low-growing, creeping perennial only a few inches tall, it produces shiny green leaves arising from the stem. In their shape, the leaves resemble those of the common garden geranium (*Pelargonium* spp.). Clusters of small white flowers (about one-eighth of an inch in diameter) nestle in the leaf axils. American marsh pennywort favors open, wet areas such as seeps, open forests, and marshes.

Panax quinquefolius L.,
American ginseng (EV) (Araliaceae)

American ginseng occurs in rich, moist hardwood forests. Flowers are about one-half of an inch in diameter, greenish-white, in a round cluster that arises from the center of the whorled leaves and at about the same height. Flowering occurs in midsummer. Its bright red fruit clusters stand out in the early autumn forest. Leaves consist of three to, more commonly, five leaflets (*quinque*, "five"; *folius*, "leaves"), which turn yellow in the fall. Roots are stubby but have long and branching smaller appendages that give them the appearance of a human torso.

American ginseng is state-protected and is highly vulnerable to overharvesting because of its putative health and high economic values.

Panax trifolius L., dwarf ginseng (Araliaceae)

Dwarf ginseng occurs in rich, moist forests, often in seep areas or on streambanks. Flowers are one-eighth of an inch in diameter, white, and in a rounded cluster at the end of the stalk, held high above the leaves in mid- to late spring. The fruit is a cluster of yellow berries. Leaves consist of three (*tri*, "three"; *folius*, "leaves") to five finely toothed leaflets and are in whorls of three. This species is not presently collected for the herbal trade, although it was used historically for numerous medicinal purposes.

Maianthemum canadense Desf.,
Canada mayflower (Asparagaceae)

Canada mayflower is one of the most ubiquitous plants of moist, upland hardwood and conifer forests throughout the Adirondacks. It also occurs on elevated microtopography in forested wetlands. The genus name comes from *Maius*, for "May," and *anthemon*, for "flower." Twelve to twenty-five densely packed white flowers occur in late spring along a terminal, short, unbranched stalk (raceme) that rises above the pair of heart-shaped leaves. The fruit is initially a green-dotted round berry that is deep red at maturity. Berries are eaten by ruffed grouse, chipmunks, and white-footed mice.

Nonflowering plants have a single leaf. This species usually occurs in dense carpets, spreading by rhizomes. It is the shortest of the four *Maianthemum* species in the Adirondacks, reaching only six inches in height.

Maianthemum racemosum L., false Solomon's seal (Asparagaceae)

False Solomon's seal occurs in rich, moderately moist, upland deciduous forests. Its flowers are white, in a branched and pyramidal terminal cluster, in mid- to late spring. Flowers are pollinated by bees, flies, and beetles. The fruit is a berry, initially green and purple dotted, becoming red with purple spots when mature, and is eaten by woodland birds and white-footed mice. Total height of the plant can reach about thirty inches. The stem is distinctly zigzag in appearance with alternating leaves, and like all *Maianthemum* species, the stem is unbranched. Leaves are stalkless and not clasping.

This species is listed as *Smilacina racemosa* in many guides. *Racemosa*, referring to the type of flower cluster, suggests that flowers are arranged in a nonbranching raceme, but the flowers are actually in a branched cluster, or panicle.

Maianthemum stellatum (L.) Link, starry Solomon's seal (Asparagaceae)

Starry Solomon's seal occurs in open or closed canopy wetlands that often are influenced by calcium-rich groundwater, as well as along moist shorelines and in other sandy substrates. It has the largest flowers (to one-third of an inch in diameter) of the four Adirondack *Maianthemum* species, but there are fewer of them in the unbranched, terminal flower clusters. Blooming in late spring, individual flowers are white and do indeed look like stars (*stellatum*, "starry"). The fruit is a green- and black-ish-striped berry, turning dark red at maturity, that is eaten by woodland songbirds. Leaves of this species appear pleated. Similar to all *Maianthemum* species, the stem is unbranched, and this species often occurs in patches, spreading by rhizomes. Total plant height is typically one to two feet.

The genus of this species is listed as *Smilacina* in many guides.

Maianthemum trifolium (L.) Sloboda, three-leaved Solomon's seal (Asparagaceae)

Three-leaved Solomon's seal occurs in the most distinct habitat of all Adirondack *Maianthemum* species, namely, open or closed acidic peatlands, typically emerging from a bed of *Sphagnum* moss. It is the only *Maianthemum* species in the Adirondacks that is a wetland obligate, that is, it does not occur outside of wetlands. *Maianthemum canadense* can be common in forested wetlands but only on elevated microtopography.

As the common name suggests, the plant usually has three alternate, oblong to lancelike leaves along the flowering stem, with an unbranched, terminal flowering stalk that holds three to eight white flowers in late spring. The fruit is a dark red berry. Total plant height is greater than *M. canadense* but generally shorter than *M. racemosum* and *M. stellatum*. The three-leaved Solomon's seal often forms such dense patches that many leaves from adjacent plants overlap, smothering all other herbaceous species.

The genus of this species has been listed as *Smilacina* in many guides.

Polygonatum biflorum (Walter) Elliott,
Solomon's seal (Asparagaceae)

Solomon's seal occurs in moderately moist to dry upland hardwood forests, streamsides, thickets, and forest edges. Blooming in spring, the tubular, greenish-white flowers with six lobes are solitary or in pairs or threes at leaf nodes along an arching stem that is usually from one to five feet long, depending on the variety (see below). The arching stem has been likened to the keel of a ship, giving rise to the saying: "Solomon's seal, to be real, must have flowers, on the keel," which distinguishes true Solomon's seals from the false Solomon's seals that have terminal clusters of flowers. Flowers are pollinated by a variety of bees, and ruby-throated hummingbirds will feed on its nectar. The distinctly veined leaves are stalkless. The fruit is a dark bluish-black berry. Scars left on the jointed rootstock from each year's stem are said to resemble the official seal of King Solomon. Its starchy rootstocks were once a source of food for Native Americans and early settlers.

Two varieties of this species are recognized: *P. biflorum* var. *biflorum* and *P. biflorum* var. *commutatum* (Schult.f.) Morong. The common names of these varieties indicate their basic difference—small and large Solomon's seal, respectively. The bowed stems of large Solomon's seal can reach over six feet in length.

Polygonatum pubescens (Willd.) Pursh, hairy Solomon's seal, is similar to *P. biflorum* in appearance but has minute hairs along the smaller veins of the short, stalked leaves. Their habitats are similar, except *P. pubescens* can be more common in rich, moderately moist forests.

Achillea millefolium L., yarrow (Asteraceae)

Yarrow is one of the most common summer-into-fall-blooming species of roadsides and old fields in the Adirondacks. It is abundant in waste places. Although its nativity is uncertain, it is widely naturalized in the United States and generally not regarded as invasive in New York. The individual flower heads are about one-fourth of an inch in diameter and consist of four to six white, notched ray flowers that surround the light yellow disk flowers. Numerous flowers comprise each terminal, forming dense, flat or rounded heads about two to four inches wide that are held high above the fernlike, aromatic leaves (*millefolium*, "thousand-leaved"). Dried flower heads can be used in winter floral arrangements. Plants can reach about three feet in height and typically form patches because of their ability to spread by rhizomes.

The genus name comes from the belief that Achilles used a plant of this genus to stop bleeding of soldiers' wounds during the Trojan War. Yarrow tea was made historically to treat colds and fever. The medical uses of yarrow are the subject of much contemporary research. Cultivars of various colors of the European variety have been selected for garden planting.

Ageratina altissima (L.) R. M. King & H. Rob.,
white snakeroot (Asteraceae)

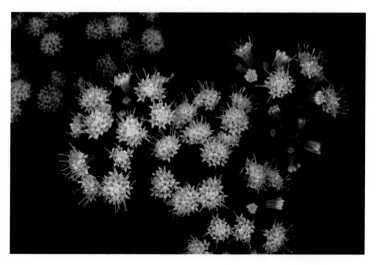

White snakeroot can be common in dry to moderately moist upland forests, especially in disturbed stands and on rocky soil, reaching up to five feet tall, although typically about half that height. It spreads locally by rhizomes, as well as seeds. Leaves are broadly oval to heart-shaped, and coarsely toothed. Numerous flower heads, one-half inch in diameter, and consisting of ten to thirty bright white disk flowers, form flat to round clusters about four to six inches wide from late summer into fall. Ray flowers are absent.

People who drank milk from cows that ate large amounts of this species often developed "milk sickness." The plant contains a toxin, tremetol, that contaminates the milk and produces severe ketoacidosis. Such milk is believed to have caused the death of Abraham Lincoln's mother. The common name, snakeroot, comes from the belief of Native Americans and early pioneers that the rhizomes of this plant could counter snakebites.

Listed in many guides as *Eupatorium rugosum*; *altissima* means "tallest" or "highest."

Anaphalis margaritacea (L.) Benth. & Hook. f., pearly everlasting (Asteraceae)

Pearly everlasting is a native species of dry, open areas including forest edges, old fields, and roadsides, often occurring in dense patches. Flower heads mature in mid- to late summer; are less than one-half of an inch in diameter; and are lemon-yellow, surrounded by white, papery petal-like bracts that persist after drying and can be used for winter flower arrangements. Ray flowers are lacking. Leaves are narrow, grayish-green on top, white-woolly beneath, and aromatic; stems are densely covered with bluish-silvery hairs. Plants can reach three feet in height.

Antennaria neglecta Greene,
field pussytoes (Asteraceae)

Antennaria plantaginifolia

Four species of *Antennaria*, or pussytoes, occur within the Adirondacks, all in similar open forests, forest edges, old fields, roadsides, or other thin soil or sandy habitats. All are native. *Antennaria neglecta* often forms dense mats of basal leaves, and the plants reach about one foot in height. Basal leaves are spoon-shaped, densely white hairy below, and arranged in a rosette. Flower heads open in mid- to late spring, are about one-half of an inch in diameter, brushlike, composed of disk flowers only, and in tight, terminal clusters. Plants are strictly unisexual, bearing only either male (staminate) or female (pistillate) flowers.

Antennaria refers to the resemblance of the male flowers to insect antennae. The common name of this genus refers to the tight, fuzzy flower heads resembling paws of kittens. *Antennaria plantaginifolia* has the longest flowering heads of any of the Adirondack species.

Dipsacus fullonum L., wild teasel (Asteraceae)

Listed in some guides as *D. sylvestris*, wild teasel is a biennial naturalized from Europe along roadsides, in old fields and waste places, and is absent from much of the Adirondacks. Pale purple to white tubular flowers are densely packed in a spiny, thimble-shaped head that is one to two inches tall. Flowering occurs from midsummer into fall. Leaves along the prickly stem appear puckered and are lance-shaped, with bases that surround the stem. Plants can reach seven feet in height.

Fullonum refers to the use of the dry flower heads for rubbing (by fullers) of woven woolen fabric. Dried flower heads are used in dried plant arrangements. It is regarded as a nuisance weed in some regions.

Doellingeria umbellata (Mill.) Nees, flat-topped white-aster (Asteraceae)

The flat-topped aster occurs in moist to wet meadows and other open areas, reaching seven feet tall. Flower heads are about three-quarters of an inch in diameter, each with seven to fourteen rays that encircle the greenish-yellow to yellow disk flowers that turn grayish. Many flower heads are held in flat-topped (sometimes dome-shaped), branched clusters three to twelve inches wide from late summer into fall. The species epithet, *umbellata*, refers to these umbrella-like flower clusters. Leaves are lance-shaped, with prominent veins; leaf margins lack teeth. Plants are unbranched except for the flower clusters.

The genus is named after Ignaz Döllinger (1770–1841), a German botanist.

Erigeron philadelphicus L., fleabane (Asteraceae)

Erigeron strigosus

Erigeron pulchellus

Fleabane is one of about six native *Erigeron* species in the Adirondacks, occurring in moist open forests, forest edges, old fields, roadsides, and other disturbed areas. Flower heads appear in midspring through summer, about three-quarters of an inch in diameter, and consist of one hundred to three hundred pink to white threadlike rays that surround the flat, yellow disk. There are usually ten or more flower heads per cluster. Stem leaves are clasping. Plants can reach about thirty inches in height. The small daisy fleabane, *Erigeron strigosus*

Muhl. ex Willd., is similar but with fewer rays and nonclasping stem leaves.

Dried plants were once burned as a fumigant to rid homes of fleas; crushed, the fresh plants produced a body lotion.

Robin's plantain, *Erigeron pulchellus* Michx., also occurs in the Adirondacks, in the same habitats as the previous species, as well as in rich, moist hardwood forests. Flower heads are up to one inch wide and consist of fifty to one hundred pale purple to pale pink rays; disks are yellow and flat. Flowering is in mid- to late spring; there is only one or a few flower heads per cluster. Its leaves are mostly basal. Plants are clonal, forming large patches that can reach about eighteen inches in height.

Eupatorium perfoliatum L., boneset (Asteraceae)

Boneset is a sunny wetland species, occurring in shallow emergent marshes and along streambanks, reaching three to four feet in height. Each flower head has nine to twenty-three white flowers in a flat to round cluster from midsummer until fall. Ray flowers are absent. Flowers attract a diversity of insects, including many kinds of bees, flies, wasps, butterflies, and beetles. Boneset is also known as "throughwort" because the densely hairy stem appears to grow through the pairs of opposite, sharply toothed, lance-shaped leaves at each stem node.

Leaves wrapped in bandages around broken bones were believed to mend them. Another explanation for the common name is that the leaves of this species, which contain the glucoside eupatorin, were made into a tonic to treat bone-breaking fever. This was one of the plants most commonly used by early pioneers for medicinal purposes.

Eurybia divaricata (L.) G. L. Nesom, white wood aster (Asteraceae)

The white wood aster is common in dry, open woods and mature forests of deciduous tree species. The one-inch diameter flower heads of about five to ten white rays and many yellow disk flowers that fade to red form flat-topped, terminal clusters from late summer to early fall. Leaves are heart-shaped, stalked, and coarsely toothed along the margin. Stems are often zigzag, and plants are usually two feet in height, but can range from one to three feet.

Leucanthemum vulgare L., oxeye daisy (Asteraceae)

Oxeye daisy is one of the most readily recognized summer wildflowers and is often included in commercial wildflower mixes. Although the species is native to Europe, it has been naturalized throughout the United States along roadsides and in old fields, waste places, and other disturbed areas. It is considered a noxious weed in some western states.

The terminal flower heads are about two inches in diameter and consist of fifteen to thirty-five white rays around a large, central yellow disk that is depressed in the center. Leaves are dark green, lobed or toothed. Plants reach about two feet in height.

Cows that graze on this plant produce an unpleasant-tasting milk. But it is used as a human food in salads. Oxeye daisy is listed as *Chrysanthemum leucanthemum* L. in many guides. This species is one of the parents of the popular Shasta daisy of gardens.

Nabalus altissimus (L.) Hook.,
tall rattlesnake root (Asteraceae)

Nabalus boottii   *Nabalus trifoliolatus*

Tall rattlesnake root, also known as "tall white lettuce," occurs in dry to moist woods of hardwood tree species. Flower heads droop, are about one-third of an inch wide and one-half of an inch long, consist of yellowish-white ray flowers, and occur in relatively tight clusters of about five. Lower to mid-leaves are three- to five-lobed and up to eight inches long.

The white rattlesnake root, *Nabalus albus* (L.) Hook., occurs in dry to moderately moist upland hardwood forests, especially in disturbed areas. Flower heads droop, are about one-half of an inch wide, consist of pink to creamy ray flowers, and are in clusters of about ten.

The three-leaved rattlesnake root, *Nabalus trifoliolatus* Cass., also known as "gall-of-the-earth," occurs in similar conditions and within the alpine zone of the Adirondacks. Flower heads droop and occur in wide-spreading clusters. The common name reflects the intensely bitter taste of aerial plant parts.

Another *Nabalus* species in the Adirondacks is restricted to the alpine zone—Boott's rattlesnake root, *N. boottii* DC. This species, named after J. W. Boott, who discovered it in 1829 in the White Mountains, has triangular to heart-shaped leaves. Plants are less than one foot in height. It is one of the rarest plants in New York, having an S1 ranking.

All of the Adirondack *Nabalus* species, except *N. boottii*, are rather tall, reaching about four feet in height (except *N. trifoliolatus* in the alpine zone is much shorter). The leaves and stems of all *Nabalus* species exude a milky sap when torn. In many guides, this genus is listed as *Prenanthes*, and "lettuce" is often part of its common names. These greens are not used for salad, as are the leaves from the commercial lettuce, *Lactuca sativa* L.

Oclemena acuminata (Michx.) Greene, whorled wood aster (Asteraceae)

The whorled wood aster is one of the more common wildflowers in moist Adirondack hardwood and hardwood conifer forests, and occurs on elevated microtopography in swamps. Flower heads are about one inch in diameter and consist of about ten to twenty-one white (to pink) ray flowers surrounding yellow disk flowers that later turn to a reddish shade. Ray flowers are not straight like those of other asters. Heads are often sparse, in flat-topped clusters. Flowering occurs from midsummer to early fall. Leaves, which are in whorls, are coarsely toothed, lance-shaped, and sharply pointed; upper leaves are larger than the lower ones. Plant parts are hairy and sticky. Plants reach about two feet in height.

Solidago bicolor L., silver rod (Asteraceae)

Silver rod, also known as "white goldenrod," thrives on rocky, dry substrates in full sun to partial shade in open forests. Flower heads consist of seven to fourteen creamy white rays and yellowish bracts along a narrow, sparsely branched, spiked flower stalk. Flowering occurs from late summer into October. Leaves are toothed, elliptical in shape, and hairy on both sides, becoming smaller toward the top end of the stem. Silver rod is quite variable in height, ranging from one-half foot tall to over three feet. It is sometimes confused with the hairy goldenrod, *S. hispida*, but the later species has yellow ray flowers.

Symphyotrichum boreale (Torr. & Gray) Á. Löve & D. Löve, northern bog aster (Asteraceae)

The northern bog aster occurs in fens and openings in cedar swamps. Flower heads are about one inch in diameter and consist of twenty to fifty white to pale blue or pinkish ray flowers around the central disk flowers. It blooms from midsummer to fall. Leaves are very narrow. Plants can reach about thirty inches in height.

Symphyotrichum lanceolatum (Willd.) G. L. Nesom, lance-leaved aster (Asteraceae)

The lance-leaved aster occurs in fields and ditches, along roadsides and streambanks, and in swamps. Flower heads are up to one inch wide and are composed of twenty to forty white to pale blue ray flowers that surround the central yellow disk flowers. Some plants will have over one hundred flower heads. Leaves are lancelike, hairless, usually toothed, and with a wedge-shaped base. This aster often occurs in dense, large patches, spreading by rhizomes. Plants can reach five feet in height.

Symphyotrichum lateriflorum (L.) Á. Löve & D. Löve, calico aster (Asteraceae)

The calico aster occurs in fields, open woodlands, and swamps, and along roadsides and streambanks. Flower heads are about one-third to one-half of an inch diameter and consist of nine to fourteen white ray flowers that surround the yellow disk flowers, which later turn reddish-purple. Heads are arranged in loose clusters that are mostly on one side of the widely spreading flower cluster. The common name refers to the multiple shades of color exhibited by the central disk flowers. Leaves are lance-shaped and hairy only along the lower midrib. The plants bloom from late summer into fall and can reach five feet tall.

Symphyotrichum pilosum (Willd.) G. L. Nesom, frostweed aster (Asteraceae)

The frostweed aster occurs in disturbed areas, including road-sides, old fields, pastures, and waste places. Flower heads are one-half of an inch in diameter and consist of sixteen to thirty-five ray flowers that surround the central, yellow disk flowers. Flower clusters are very open and widely branching. Leaves are narrow and sharp-pointed, with the lower leaves often falling off later during the growing season. It reaches five feet in height.

Podophyllum peltatum L., may apple (Berberidaceae)

Little introduction is needed for this well-known, common woodland plant. It arises from an extensive creeping rhizome that can extend hundreds of square feet. Although it is a forest dweller, populations of may apple may persist for many years after timber removal. It is common along shaded roadsides, often in association with white trillium.

No other native plant has the large, umbrella-like leaves with four to six lobes, each broadly forked at the tip. Populations of may apples are often infected with a rust fungus that produces bright orange lesions on the undersurface of the leaves. This may apple rust (*Allodus podophylli* Schwein) apparently does no permanent harm to the host.

There are usually two leaves on the stem with a single, large white flower in the fork at the base of the two leaves. The five to seven petals are thick and waxy. Pollen is released in may apple, as in other flowers in the same family, by two flaps (rather than slits) in the anther. The fruit may be difficult to see under the large leaves, but it is placed at just the right height for box turtles, which (south of the Adirondacks) eat the fruit and spread the seed.

Pharmacological studies of genes in may apple have revealed products with potential efficacy for cancer treatments. These and other compounds in the plant make it toxic.

Hydrophyllum virginianum L., Virginia waterleaf (Boraginaceae)

The common name "waterleaf" is thought to refer to the white splotches on the leaves, found mainly on those that were the earliest to appear. These are pinnately divided (with opposite lobes, like the structure of a feather, *pinna* in Latin). The stem is usually less than two feet tall. Clusters of flowers are produced in late spring. They range from almost white, to pinkish, to light purple. Bell-shaped, the flowers have five lobes. Because so many flowers appear at one time on a plant, a large population of waterleaf is a beautiful sight.

It is a spring ephemeral, although it flowers later than many of its guild. Waterleaf grows in moderately moist, upland forests but can also occur in floodplains.

Cardamine diphylla (Michx.) Wood, two-leaved toothwort (Brassicaceae)

Cardamine concatenata

A true spring ephemeral, this toothwort appears with a guild of other harbingers of spring like bloodroot, Dutchman's breeches, and hepatica. This is a perennial of rich, moist, upland forests. As the specific epithet implies, there are two main leaves opposite one another on the stem. The leaves are three-parted. Flowers are white with four distinct petals.

The related cut-leaved toothwort, *Cardamine concatenata* (Michx.) O. Schwartz, is less frequent. It has leaves that are also divided into parts, usually three to five linear segments. Flowers are white and resemble those of the two-leaved toothwort. Cut-leaved toothwort leaves tend to be two- rather than three-lobed.

There are several other species of *Cardamine* in the Adirondacks. Species are separated on the basis of leaf shape, petal length, and other technical characteristics.

Nasturtium officinale W. T. Aiton, water cress (Brassicaceae)

Like dandelion, water cress was likely introduced from Europe as a food plant and has now spread widely. It is grown commercially for the young leaves that are used in salads and vegetable drinks because of their pungent taste of mustard. Older plants tend to be bitter but are still edible.

Water cress is not common in the Adirondacks but can form large colonies in springs, seepages, margins of streams, and the mucky soils of drainage channels. This annual has alternate, deeply lobed leaves. Flowers are white and have the typical four petals in a cruciform pattern that define members of the mustard family. Flowering occurs in midsummer, yielding a long, narrow pod with round seeds. Water cress is easy to grow from seed, which is often sold for producing sprouts. The plant is very nutritious, rich in vitamins A and C.

It is also known as *Rorippa nasturtium-aquaticum* (L.) Karsten. It is no relation to the common garden nasturtium, *Tropaeolum majus* L. (Tropaeolaceae).

Campanula aparinoides Pursh, marsh bellflower (Campanulaceae)

This is a delicate, weak-stemmed perennial up to two feet tall but seldom growing erect. Rather, it falls onto surrounding vegetation for support and for that reason is difficult to see unless in flower. Leaves are alternate, narrow, about one-and-a-half inches long, and have recurved hairs on their margins that allow them to grip surrounding vegetation. That is why this species was named after another, unrelated, plant—*Galium aparine*—which is distinguished by its hairs that allow it to cleave to other plants (and clothes), giving it the common name of "cleavers."

As its name implies, the flowers are bell-shaped (*campana* in Latin), less than a quarter of an inch long, usually light blue to white. It grows in marshes, seeps, wet meadows, and fens.

Mononeuria groenlandica (Retz.) Dillenb. and Kadereit, mountain sandwort (Caryophyllaceae)

A true boreal plant found as far north as Labrador and Greenland, mountain sandwort grows in the Adirondacks only in the alpine tundra of the High Peaks. Here it is associated with the exclusive guild of other tundra plants with which it shares some features, such as being mat-forming and having small (opposite) succulent leaves. It favors granitic rocks.

White, five-parted flowers appear from midsummer until fall and are visited by flies. These insects, unlike the bumblebees that can also be pollinators, are active during the cloudy and windy days frequent on the High Peaks. Like most members of the pink family, Caryophyllaceae, each petal is notched at the tip, as if it had been cut with pinking shears—hence the common name of the family.

Silene latifolia Poir., white campion (Caryophyllaceae)

White flowers appear from summer to fall and are only fully open at night, like many other white flowers that attract moths. White campion is unusual because each plant is unisexual, either producing all female (pistillate) or all male (staminate) flowers. The attractive blossoms are about an inch across and displayed at the top of the plants; plants are usually about two feet tall. The five petals are each incised, giving the appearance of ten petals. The calyx at the base of the petals is swollen, which creates a bulb-like appearance. Pistillate flowers have five white styles; staminate flowers have ten stamens. Both the styles and stamens extend beyond the throat of the flower. The leaves are opposite and, like the stems, are covered with soft hairs. These hairs can be clammy to the touch and sometimes trap small insects—the source of the name "fly campion."

White campion contains saponins (chemicals that produce a soapy lather) that render the plant toxic to grazers (as well as humans). Although introduced from Europe and weedy, this annual or biennial adds color to roadsides, meadows, and disturbed sites.

Silene vulgaris (Moench) Garcke, bladder campion (Caryophyllaceae)

Bladder campion is common in the Adirondacks in old fields, roadsides, and other open disturbed areas. A European introduction, perhaps as a garden plant, bladder campion is now widespread through the northern United States and adjacent Canada.

A perennial with opposite leaves, bladder campion grows to about two feet tall and produces white flowers from midsummer to fall. Leaves are opposite and often clasp the stem. There are five petals, but they are so deeply incised that they appear to number ten. The most distinctive feature of this species is the inflated calyx, which has an intricate reticulate venation pattern, often pink in color.

Parnassia glauca Raf., grass-of-Parnassus (Celastraceae)

Parnassus is a mountain in central Greece, sacred to the Greek gods. The Greek botanist Dioscorides (circa AD 90) named a plant from this mountain after the limestone massif. It is not a grass and does not resemble a grass.

A cluster of leaves with more or less ovate blades are present at the base of this perennial of open fens. From the center of these leaves one or more flower stems arise bearing white flowers about one and a half inches wide. Most striking are the green veins embossed in the petals, a feature unique in our flora. These markings are thought to be guides for pollinators.

Cuscuta gronovii Roem. and Schult., common dodder (Convolvulaceae)

A sizable population of common dodder looks like spaghetti thrown over streamside plants. This dodder favors wet areas where the hard seeds can be scarified by the action of water scraping them over rock and sand.

Remarkably reduced for its parasitic existence, dodder lacks roots and leaves. The plant consists of nothing but stem and flowers. All nutrition is obtained from the hosts, which are invaded by a specialized organ called the haustorium that produces a peglike outgrowth that bores into the host to connect it with food-carrying tissue. Haustoria become evident when the parasite is peeled off its hosts. Common dodder is not host specific and will attach to a diversity of herbaceous and woody plants. Flowers are white, five-parted, and produce a dry, papery fruit with three or four seeds.

An attractive yellow dye can be made from the plant. It is named after Jan Frederik Gronovius, 1690–1762, a teacher of Linnaeus and the author of the first flora of the New World, *Flora Virginica* (1739).

Cornus canadensis L., bunchberry,
dwarf dogwood (Cornaceae)

The beauty in both flowering and fruiting of this common, widespread component of the northern hardwood and boreal forests is reflected in its two common names. Its relation to the widespread flowering dogwood, *Cornus florida*, is evident when it is in flower because of the large white bracts. The relatively large (in comparison to the rest of the plant) "flower" is actually a collection of tiny, greenish-yellow flowers surrounded by four showy white bracts. In midsummer, bright red fruits are produced, giving it the other common name of "bunchberry."

The flowers are small but powerful. Each flower produces a device akin to a trebuchet, the fulcrum war machine of the Middle Ages that could fling burning oil or rocks long distances to inflict damage on the enemy. The projectile of the bunchberry is pollen: it leaves the flower faster than a bullet from the barrel of a rifle—clocked at 0.04 milliseconds, the fastest recorded movement of any plant. This armament shoots the pollen into the air stream for travel to a neighboring plant—an important device for a plant the stature of bunchberry, which is only a few inches tall.

The red fruits are edible, slightly sweet but usually dry and insipid. They add bright color to a summer salad.

Drosera intermedia Hayne, spatulate-leaved sundew (Droseraceae)

Drosera rotundifolia L., round-leaved sundew (Droseraceae)

These are small, easily overlooked plants. But sundews can provide a burst of color when light strikes the large, stalked, sticky glands covering the leaves. Sundews are part of a guild of plants augmenting their nutrition by catching and digesting animals—in the case of sundews, usually insects. Drawn by the luscious-looking glandular hairs, insects land on the leaf only to find that they are unable to extricate themselves. Inexorably, the gland bends over the insect, forcing it to the surface of the small leaf, where additional digestive glands are located. Then, imperceptibly, the leaf folds, covering the hapless victim. These same meat-eating leaves form a compact, overwintering structure in some species.

Meanwhile, above the carnage the plant pursues its sex life. It produces white, five-parted flowers on stalks two to three inches tall throughout the summer. Capsules with many tiny seeds soon appear.

Sundews grow in nutrient-poor areas because their carnivory augments their nitrogen budget. These two species often grow near each other but round-leaved sundew is found in slightly drier conditions, and spatulate-leaved sundew is usually found on wetter sites, like saturated mudflats. They each are easily recognized by the shape of the leaf. As the species names indicate, one has round leaves; the other, spoon-shaped. These sundews often form sizable populations in bogs and marshes as well as on moist margins of ponds and streams.

Chimaphila umbellata (L.) Barton, pipsissewa (Ericaceae)

"Pipsissewa" is a Native American name for this charming evergreen plant that is distributed around the globe. Typically six to twelve inches tall, it has toothed, shiny leaves. The lustrous green color of the plant belies an important aspect of

its physiology. Pipsissewa, like other members of the blueberry family (Ericaceae) to which it belongs, is partially dependent
on fungi that are associated with tree roots. These fungi, many
of which are mushrooms, convey nutrition from the food-pro-
ducing trees to the plant. Although pipsissewa contains chlorophyll and can produce some of its own food, this heavy de-
pendence on mycorrhizae may be an adaptation to growing in
shade, where less light is available for photosynthesis.

Flowers are attractive, with five pink or white petals, and
are produced on recurved stalks. Pollen is shed by pores in the
anthers. The single, large stigma glistens at maturity. Each
fruit, a dry capsule, contains thousands of tiny dustlike seeds,
typical of plants with strong mycorrhizal relationships.

The leaves have a pleasant flavor, owing in part to the compound hydroquinone. They are dried and used to make a tea.
Large quantities, however, should be avoided because of putative carcinogenic properties of hydroquinones.

Gaultheria hispidula (L.) Muhl. ex Bigelow, snowberry (Ericaceae)

Snowberry is often the dominant groundcover on shaded hummocks and decaying
stumps in conifer swamps. It
has a trailing, vinelike habit much like cranberries in
the same family. It has nearly

round, evergreen leaves that are smaller in diameter than a fingernail on one's pinky finger; both leaf surfaces have distinct, scattered brown hairs that are especially prominent on the leaf underside. Its snow-white oblong fruit has a wintergreen odor when crushed.

Gaultheria procumbens L., wintergreen (Ericaceae)

One of the best known and easiest to identify of Adirondack plants, wintergreen is common throughout the region in bogs, cut-over forests, or any acidic soil and is one of the guild of northern forest plants that extends across the eastern United States and adjacent Canada. It does not thrive in areas with normal concentrations of calcium in the soil. The shiny, dark evergreen leaves are borne on a narrow, woody rhizome. Urn-shaped flowers appear in the early summer and produce the wintergreen "berry," actually a specialized fruit containing the dry capsule embedded in the fleshy structure. The fruits persist into the winter, making wintergreen an attractive Christmas decoration.

This evergreen subshrub was once of economic value as the commercial source of wintergreen oil for flavorings and medical uses. All parts of the plant are edible—and potentially toxic. Small quantities are tasty, but the oil extracted from the plant, methyl salicylate, can be toxic—even fatal—at relatively small doses.

Moneses uniflora (L.) A. Gray, one-flowered wintergreen (Ericaceae)

The one-flowered wintergreen, also known as one-flowered shinleaf, occurs on moss-covered substrates, often rotting logs, in moist to wet hardwood and hardwood conifer forests, and on elevated microtopography in conifer swamps. It is the only shinleaf with a single flower on each plant. The plant blooms in midsummer; flowers are fragrant, nodding, white, about one-half of an inch in diameter, with five petals. The fruit is a capsule. The evergreen leaves are roundish and finely toothed, and, as in most other shinleafs, are in a basal whorl. Total plant height is about five inches.

The genus name comes from *monos* ("single"), and *esis* ("delight"), referring to the single, attractive flower.

Monotropa uniflora L., Indian pipe (Ericaceae)

Few plants garner more questions and interest than Indian pipe. No wonder. Its ghostly white appearance (sometimes tinged with pink) makes it look most un-plantlike, an alien in verdant surroundings. Yet this five- or six-inch high plant is the most connected member of its community. Virtually all the plants around it live in association with fungi called mycorrhizae. Many mushrooms are mycorrhizal. These mushrooms associate with roots of trees and other plants, interconnecting all the plants through an underground botanical network. Food produced by green plants is used by the fungi, and they in turn deliver hard-to-obtain minerals to Indian pipe. Indian pipe exploits these relationships. This relative of blueberries and cranberries taps into the mycorrhizae of the trees to obtain all of its nutrition. It is so specialized that it has lost the ability to make its own food. No photosynthesis is necessary, so Indian pipe lacks chlorophyll.

Appearing in mid- to late summer in a diversity of habitats, Indian pipe obtained its common name from the single, downward-pointing flower that gives this pallid plant the appearance of an old-fashioned tobacco pipe. As the fruit develops, the bowl of the pipe straightens to allow the large number of tiny seeds to disperse from the capsule. Just how these seeds germinate and initiate their relationship with their host fungi remains unknown.

Pyrola americana Sweet, round-leaved shinleaf (Ericaceae)

This plant is a rhizomatous perennial with leaves at the base of the stem. Leaves are round, shiny, and evergreen. Attractive flowers are produced from May through midsummer on a terminal stalk. Flowers are white, waxy in appearance, with five petals and a distinct outward thrusting, curved style. In late summer, a flattened brown capsule with one thousand or more tiny seeds is produced. Like other members of this genus, round-leaved shinleaf is strongly mycorrhizal, so little food is stored in the seeds. It grows in dry to moderately moist upland forests, often on sandy soils.

Pyrola elliptica

Common shinleaf, *Pyrola elliptica* Nutt., is the most frequent shinleaf in the Adirondacks. It has flowers and fruits similar to those of the round-leaved shinleaf. The two can be distinguished by the habitats in which they occur and the size and texture of the leaves; those of common shinleaf have a dull surface and are longer than wide. Common shinleaf is found in moister habitats.

Eriocaulon aquaticum (Hill) Druce, northern pipewort (Eriocaulaceae)

This is a charming, easily missed plant with some of the smallest flowers of any plant in the Adirondacks. It is an aquatic found in shallow streams, intermittent pools and seepages, and ponds. In deeper water it can spend part of its life submerged.

Single flowering stems arise in the summer from a rosette of leaves about the size of a half-dollar. The stem is topped with a disk-shaped flower cluster resembling an old-fashioned hatpin. The flower head with its dozens of flowers is about an inch and a half wide. Flowers are unisexual and tiny, only a millimeter or so wide. They appear black and white because of the abundance of white, club-shaped hairs on the flower parts.

The black color comes from the modified leaves that surround the head; each flower petal also has a black dot on it.

Since pipewort has nectar glands, insect pollination is implicated. The fruit is a few-seeded capsule. Pollination biology of this widespread species has been little studied, but flies are likely pollinators as has been recorded for other species of pipewort.

Lycopus americanus Muhl., American bugleweed (Lamiaceae)

Lycopus uniflorus

This is a very common perennial found in a wide variety of wetlands. The plant is about twelve inches high, with the stems often bending and rooting in the damp soil. Flowers are very small and produced in crowded clusters in the axils of the upper leaves. The flowers are bell-shaped and white. Leaves are strongly toothed and aromatic.

Another species, *Lycopus uniflorus* Michx. (northern bugleweed) is also widespread in similar habitats in the Adirondacks. The two species are very similar in appearance.

However, northern bugleweed often has purplish leaves. Both species are rhizomatous, but only American bugleweed has small tubers at the tips of the rhizomes. These are starchy, edible, and unremarkable in taste.

Streptopus amplexifolius (L.) DC., clasping-leaved twisted stalk (Liliaceae)

Clasping-leaved twisted stalk, also known as "white mandarin," occurs in moist, rich upland forests, flowering from late spring into early summer. Its flowers, on slender stalks, are white to greenish and occur singly or in pairs. Leaves are oblong and clasp the branched stem, which can reach about one to three feet tall.

The nodes of the stem are smooth, versus the finely hairy leaf nodes of rose twisted stalk. The fruit of this species is an elliptical, shiny, bright red berry. There are conflicting reports about the toxicity of this and related species. It resembles green hellebore, which is a documented highly toxic plant.

Malva moschata L., musk mallow (Malvaceae)

Musk mallow was introduced from Europe and northern Africa and is now naturalized in much of North America, occurring in the Adirondacks along roadsides and in old fields, pastures, and other disturbed soils. Flowers are musk-scented, one and one-half to three inches in diameter, and consist of five white or pink triangular-shaped notched petals, in terminal clusters flowering from summer until fall. Leaves along the stem are deeply dissected and are also musk-scented when crushed. Plants are bushy and can reach about three feet in height. The leaves and flowers are edible and can be added to salads.

Trillium cernuum L.,
nodding trillium (EV) (Melanthiaceae)

Of the four *Trillium* species in the Adirondacks, *T. cernuum* occurs in the most moist, even wet conditions, including lowland forests, as well as in upland forests of mixed hardwoods and conifers. It blooms about one month later than the purple and white trilliums, flowering in late spring, although a casual hiker would likely not even know when it is in bloom. *Cernuum*, means "nodding" or "drooping," referring to the flowers that are held below the plant's three, nearly overlapping leaves. Nodding trillium has a white ovary, and its fruit is a dark red, oval berry.

Trillium grandiflorum (Michx.) Salisb., white trillium (EV) (Melanthiaceae)

The white trillium occurs in rich, moderately moist forests typically dominated by sugar maple (*Acer saccharum*) and American beech (*Fagus grandifolia*), and it is often found in the open along roadsides adjacent to deciduous woodlands. It blooms in early spring, its white flowers turning pinkish-purple before withering. The fruit of white trillium is a round but faintly six-angled berry.

In stands where white and purple trillium occur together, the white trillium blooms as the purple trillium flowers are fading. The species name, *grandiflorum*, suggests a much larger flower for this species than the other *Trillium* species. It is indeed larger, but the size difference is not usually very significant. However, when massed together at peak bloom, few wildflowers at any time of year rival the beauty of this species. Flowering intensity does diminish in the deepest shade. It is the provincial flower of Ontario, Canada.

The white trillium is one of the easiest of *Trillium* species to cultivate in the garden, increasing in stem and flower number under good conditions. Double-flowered forms are especially attractive, as are those that have pink flowers upon opening.

Trillium undulatum Willd.,
painted trillium (EV) (Melanthiaceae)

The painted trillium is rather common in Adirondack forests, especially those on acidic, humus-rich, upland sites dominated by eastern hemlock (*Tsuga canadensis*), red spruce (*Picea rubens*), eastern white pine (*Pinus strobus*), and drier-site hardwoods like red maple (*Acer rubrum*). In the Adirondacks, it seems always to be associated with pink lady's slipper (*Cypripedium acaule*), and it blooms about the same time. Blooming weeks later than the purple and white trilliums, the painted trillium would stand out at any time with its striking, dark reddish, diffusely V-shaped markings that converge at the center of each flower. *Undulatum* means "undulating" or "wavy" and refers to the margins of the flower petals. Its leaves are a distinct shade of dark green, with a purplish cast as they unfold. The cylindrical, scarlet berry is held erect above the leaves.

Although the painted trillium is so common in the Adirondacks that one would assume it is easy to cultivate, this is not actually the case—unless one's garden has the acidic, deep humus, cool soil, and just the right amount of light in which this species naturally thrives.

Menyanthes trifoliata L., buckbean (Menyanthaceae)

Buckbean flowers at the same time as marsh marigold and pussy willows. The five white petals are covered with long hairs, giving the flowers a lacy appearance, perhaps to block small non-pollinating insects from collecting pollen or nectar. These hairs, however, are no barrier to the larger and stronger bumblebees that are the most common pollinators.

Often growing in standing water of ditches, bogs, and fens, buckbean is found around the globe in northern regions. As the specific epithet describes, leaves are three-parted, resembling the leaves of many members of the unrelated bean family. The seed, hard and shining, resembles a bean. However, buckbean should not be eaten, as there are reports of serious nausea from ingesting the roots. Despite this, some wild edible plant guides include it as a food that requires considerable preparation of boiling and repeated rinsing.

Nymphaea odorata Aiton, fragrant white water lily (Nymphaeaceae)

The most familiar aquatic plant of lakes in the Adirondacks is the white water lily. Large, white, fragrant flowers are numerous throughout the growing season, adorning the margins of lakes and ponds in water six feet deep or less.

Leaves are round with a narrow V-shaped slit that prevents damage to the leaf when it is stressed by waves and wind. At each side of the base of the slit is a distinct lobe, the auricle, so named because it resembles an ear lobe. Fragrant white water lily leaves resemble those of the common yellow pond lily but the auricles enable recognition of *Nymphaea*, a worldwide genus. Yellow pond lily lacks auricles on the leaves. Leaves of the fragrant white water lily arise from large, fleshy rhizomes that are a favorite food of muskrats.

Flowers are functional for two days. On the first day, the flower is functionally male when it opens—usually between 7:30 and 9:00 a.m. By 4:00 p.m., the flower closes; it reopens the next day, this time as functionally female. A diversity of insects visit the flowers; one of the most common is the introduced honeybee, drawn by the color, fragrance, and heat that the flower generates through a process known as thermogenesis.

After fertilization, the flower is drawn under water by a coiling of the flower stalk, and the round fruit develops submersed. It matures in about five weeks and then disintegrates, with each seed encased in a buoyant covering that aids in dispersal. After a while, the casing decays and the seed settles to the bottom. Floating seeds are food for waterfowl.

Circaea alpina L., dwarf enchanter's nightshade (Onagraceae)

Circaea canadensis

Any hiker in the Adirondack forests in the fall has likely had an encounter with this plant. The club-shaped, bristly fruits stick to clothes—an obvious and effective means of dispersal by animals. Growing to less than a foot tall, dwarf enchanter's nightshade is a modest plant that flowers in midsummer. The flowers are a quarter of an inch across and have only two white petals—an unusual occurrence in a family with such showy flowers such as primroses, which have five petals. The leaves are opposite with only a few pairs on the stems. Leaves are delicate, thin, and have scattered teeth on their edges.

Another species of nightshade, eastern enchanter's nightshade [*Circaea canadensis* (L.) Hill], also occurs in the Adirondacks. The two species are separated on the characteristics of their rhizomes and overall size, with eastern enchanter's nightshade being larger. As a true boreal species, dwarf enchanter's nightshade favors cooler sites in moist upland forests, while the other enchanter's nightshade, though native, can be found in disturbed sites.

Goodyera pubescens (Willd.) R. Br., downy rattlesnake plantain (EV), *Goodyera repens* (L.) R. Br., dwarf rattlesnake plantain (EV), and *Goodyera tesselata* Lodd., checkered rattlesnake plantain (EV) (Orchidaceae)

Goodyera pubescens   *Goodyera repens*

Goodyera tesselata

The three *Goodyera* species in the Adirondacks have similar evergreen and leathery leaves in a basal rosette, stalks of small white flowers that bloom during the summer, and rhizomatous habit. They typically occur in colonies, some dense and large. Of the three species, *G. pubescens* is most likely to be found in moderately moist deciduous forests, while *G. repens* and *G. tesselata* thrive in dry upland forests often dominated by evergreen tree species, especially pines (*Pinus strobus* and *P. resinosa*) and eastern hemlock (*Tsuga canadensis*). The substrate for all three species, regardless of canopy composition, tends to be acidic.

The downy rattlesnake plantain is named for the hairiness (*pubescens*) of its stalk and flowers, although this characteristic is not a clear distinction from other *Goodyera* species. It can have up to eighty flowers uniformly distributed around the flower stalk, whereas the open flowers of *G. tesselata* and *G. repens* tend to be one-sided. A study examining the environmental factors that affect flowering in *G. pubescens* demonstrates the complexity of flowering variability within and among orchid species. Extensive periods of warm, dry weather during May of the previous year explained the synchronicity (blooming at the same time) of this species (Reddoch and Reddoch 2007).

The dwarf rattlesnake plantain is smaller in stature, as the name suggests; the species name, *repens*, in botanical Latin means "creeping," a trait of all three species. The leaves of *G. repens* are smaller and less "checkered" than those of *G. tesselata*, and the maximum number of flowers per stalk is twenty. *Goodyera repens* is well known in Eurasia as well.

The leaves of *G. tesselata* have a whitish, checkered appearance; *tesselata* means "tessellated," which in turn means "checkered" or "mosaic." This species has up to forty flowers per stalk.

Goodyera is named for the English botanist John Goodyer (1592–1664). Species in this genus are among the types of orchids that have been collected for terrarium plantings because of their very attractive foliage. Collecting these species for any purpose is illegal in New York.

Platanthera blephariglottis (Willd.) Lindl., white fringed orchid (EV) (Orchidaceae)

The white fringed orchid is an often abundant species of open, acidic peatlands (bogs), nearly always occurring along with grass pink (*Calopogon tuberosus*) and rose pogonia

(*Pogonia ophioglossoides*), which start blooming about one month earlier; as these pink orchids wane, the white fringed orchid blooms from mid-July into early August. It is not unusual to find dozens of plants at some locations, typically interspersed among clumps of the pitcherplant (*Sarracenia purpurea*) and arising from lush *Sphagnum* carpets. The white fringed orchid disappears as the open bog succeeds to a black spruce–tamarack bog.

Blephariglottis means "eyelashed tongue," referring to the heavily fringed lip. The other white-flowering *Platantheras* in the Adirondacks (*P. clavellata*, *P. dilatata*, *P. macrophylla*, *P. obtusata*, *P. orbiculata*) do not have fringed flowers and appear in very different habitats (wet, disturbed areas for *P. clavellata*, open fens for *P. dilatata*, deeply shaded lowland conifer forests for *P. obtusata*, and rich upland forests for *P. macrophylla* and *P. orbiculata*). The other *Platantheras* in the Adirondacks with fringed flowers are *P. grandiflora*, *P. psycodes*, and *P. lacera*. The first two have purple flowers; the third has greenish-white to white flowers, but its blossoms are much more deeply cut than those of *P. blephariglottis*, and although it can be found in bogs, it is more likely to appear in a much broader range of open, moist habitats.

Platanthera dilatata (Pursh) Lindl. ex Beck, tall white bog orchid (EV) (Orchidaceae)

There are few stronger, more pleasant smells than that encountered upon walking into an open, rich fen during midsummer

when the tall white bog orchid is blooming and the fragrance of cloves (or vanilla) permeates the air. This is also a striking species in bloom, especially when many individuals are scattered throughout the fen. Flowering stalks can reach over three feet tall and bear up to one hundred bright, pure white flowers that look like birds in flight.

Superficially, this species can be confused with *P. huronensis*, but *P. huronensis* has greenish-white or whitish-green flowers, not the luminous, utterly white flowers of *P. dilatata*. Moreover, *dilatata*, meaning "dilated" or "widened," refers to the base of the flower's lip, which is much wider in *P. dilatata* than in *P. huronensis*. The tall white bog orchid is also widely known as "bog candles"; however, this species rarely occurs in bogs, instead favoring peatlands with a higher pH.

Platanthera macrophylla   *Platanthera orbiculata*
(Goldie) P. M. Brown,   (Pursh) Lindl., small-
large-spurred round-leaved   spurred round-leaved
orchid (EV) (Orchidaceae)   orchid (EV) (Orchidaceae)

Platanthera orbiculata

These species are very similar in appearance, with two nearly round basal leaves that are typically flopped on the ground
and stalks of about ten to thirty whitish flowers in midsummer.
Despite the species name, *macrophylla*, or "large leaf," its leaf
size does not consistently distinguish this species from the other. The best way to separate them is by the length of the flowering spur, which is less than one and one-tenth inches for *P. orbiculata* and at least that long for *P. macrophylla*. Both species
occur in rich, deeply shaded, moist upland forests of sugar maple, American beech, and eastern hemlock, but *P. orbiculata* is

more likely in conifer swamps. Where they occur together, they can hybridize, making their distinction even more difficult.

Spiranthes cernua (L.) Richard, nodding ladies' tresses (EV) (Orchidaceae)

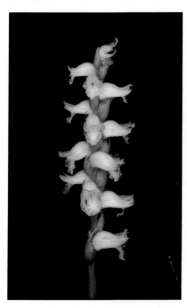

Nodding ladies' tresses is a rather common species of *Spiranthes*, occurring in open, moist to wet areas, such as meadows, ditches, old fields, borrow pits, and near springs. These habitats are often very wet in the spring and quite dry later during the growing season. It is often found in large numbers (in the hundreds to thousands). Like many of the orchid species that thrive in very sunny habitats, it disappears with shading by dense vegetation. Despite its species name, *cernua*, which means "nodding," and its common name, the flowers do not droop noticeably more in this species than they do in others of the *Spiranthes* genus. They are usually pure white and lack fragrance. The plant starts to flower in late summer, and, of all the *Spiranthes* species covered in this guide, it continues to bloom the latest through early fall—the last of all orchid flowers to be seen each growing season in the Adirondacks.

Spiranthes lacera (Raf.) Raf.,
slender ladies' tresses (EV) (Orchidaceae)

Slender ladies' tresses occurs in open dry woods, pine stands, fields, and roadsides, flowering in midsummer. Older guides refer to this highly variable species as *S. gracilis*, but *S. lacera* is now recognized as having two varieties, both in New York: *Spiranthes* var. *gracilis* (southern slender ladies' tresses) and *Spiranthes* var. *lacera* (northern slender ladies' tresses). *Spiranthes lacera* var. *lacera* is most common in the Adirondacks and is separated from *S. lacera* var. *gracilis* by looser flowers on the spike and the presence of basal leaves during flowering in var. *lacera*. *Lacera*, meaning "torn," refers to the ragged lip margin.

Spiranthes lucida (H. Eaton) Ames,
shining ladies' tresses (EV) (Orchidaceae)

Flowering in late spring to early summer (the earliest-blooming *Spiranthes* species in New York), with a deep saffron-yellow lip, shining ladies' tresses is easy to identify. This species occurs in open, wet meadows (including wet cow pastures), fens, ditches, shores of rivers, and other moist, disturbed areas. It is more frequent and abundant in areas influenced by calcium-enriched waters. *Lucida* ("shining") refers to its glossy leaves.

Spiranthes romanzoffiana Cham., hooded ladies' tresses (EV) (Orchidaceae)

Hooded ladies' tresses is the tallest and most robust *Spiranthes* in the Adirondacks, especially in open, wet meadows and rich fens, reaching nearly twenty inches in height under the best conditions. Unlike the seasonal wetlands in which *S. cernua* is commonly found, hooded ladies' tresses thrives in groundwater-fed wetlands in which the water table fluctuates little from beginning to end of the growing season. This species flowers in late summer and can occur in very large numbers, up to at least tens of thousands at one site in the Adirondacks. With up to sixty creamy-white flowers in three rows per stem, the common name refers to the sepals and petals that flare outward at the top of the flower, forming a hood. Additionally, this is the only *Spiranthes* with a fiddle-shaped lip.

Discovered on the Aleutian island of Unalaska, this species is named after Prince Nicholas Romanzoff, who financed a round-the-world expedition in 1816–17. It is geographically widespread, occurring in the mountains of the western United States to Alaska, and is even in Ireland.

Melampyrum lineare Desr., cow wheat (Orobanchaceae)

Cow wheat is a parasite that attaches to diverse hosts to obtain water and nutrients—a process driven by high rates of transpiration, requiring growth in an open, sunny area. Yet this parasite is green and can produce most of its food through photosynthesis. Haustoria, the modified roots of cow wheat that form the morphological and physiological bridge between host and parasite, are small (less than an eighth of an inch) and white. They must develop soon after germination to ensure the survival and vigor of the plant.

These diminutive plants produce small, white-and-yellow, two-lipped flowers in midsummer that soon develop into distinctive capsules containing two or three brown seeds. Some may turn black, but these will not germinate. The seeds contain an elaiosome, an outgrowth of the plant placenta that covers the seed with an oil- and nutrient-rich layer to attract ants; ants carry the seeds to their nests to feed their larvae and unwittingly disperse the plant.

Oxalis montana Raf.,
northern wood sorrel (Oxalidaceae)

Northern wood sorrel is a common small perennial of rich, moist upland forests and grows on hummocks in lowland forests. Flowers are white with conspicuous purplish-pink stripes and appear in mid- to late summer, solitary on stems that extend above the leaves. As the season progresses, typical flowers are replaced by cleistogamous flowers, small bud-shaped structures that produce seed without pollinators.

Although not obvious to the naked eye, the surface of the leaves contains altered cells that function as virtual lenses that focus light to increase the efficiency of photosynthesis for this shade-dwelling plant. These pyramid-shaped cell surfaces are evident with a 10× hand lens.

The genus name means "sour," an accurate description of the taste of the shamrock-shaped edible leaves. However, only small amounts should be ingested because of the presence of calcium oxalate, a compound that can form kidney stones in susceptible persons.

Dicentra cucullaria (L.) Bernh.,
Dutchman's breeches (Papaveraceae)

Perhaps there is no better known and more charming spring wildflower than Dutchman's breeches, with its distinct-ly shaped flowers and fernlike leaves. It is one of the guild of

Dicentra canadensis

spring ephemerals, harbingers of spring that capture the sun's energy before the forest trees leaf out.

Dutchman's breeches grows in a habitat rich with ephemerals—all well-known wildflowers, including hepatica, spring beauty, bloodroot, and more. What distinguishes Dutchman's breeches is its unique blossoms. Each of the white flowers with its yellowed tip is divided into two parts, giving the impression of an old-fashioned pair of Dutch pantaloons, hence the common name. The flowers are adapted for pollination by bumblebees that have the strength to open the flower so the nectar can be harvested and the flower pollinated. Flowers are borne hanging from a stalk picturesquely described by the Native American poet Russ Littlecreek:

I thought I missed the time of year
when droves of Dutch appear
to flaunt their frilly breeches.

The leaves are also frilly, divided into three leaflets each with narrow segments. Plants arise from small bulbs (actually tubers). Like its relative, squirrel corn [*Dicentra canadensis*

(Goldie) Walp.], the tubers of Dutchman's breeches should not
be ingested because of the presence of alkaloids. Squirrel corn is
also widespread in the Adirondacks in habitats similar to those
that favor Dutchman's breeches. Unlike Dutchman's breeches,
its flowers are heart-shaped, with a fragrance resembling hy-
acinths. The tubers of squirrel corn resemble a grain of corn.

Sanguinaria canadensis L., bloodroot (Papaveraceae)

An iconic spring wildflower widespread in Adirondack deciduous forests, bloodroot is one of the spring ephemerals—plants
that rush through their life cycles before the trees form a canopy that shuts out the light necessary for photosynthesis.

In early spring, bloodroots emerge from thick rhizomes
with usually one palmately lobed leaf per plant. One to three
flowers are borne on a stalk that rises above the leaf, usually
five to eight inches tall. Petal number is variable, from five to

ten. Like other members of its family, bloodroot has only two sepals. These often drop after the flower opens. The leaf stalk, flower stalk, and rhizome of the plant all contain a blood-red sap that gives the plant its common name.

By late spring, the fruits develop. These are long and narrow and contain two rows of dark seeds. Each seed bears a white growth (called an "aril") that is rich in protein. Ants harvest the seeds for the aril and carry them to their nests, where they can germinate with a supply of food from the ants' excrement.

Extracts of bloodroot have long been valued as medicines. Recent research has centered on the compound sanguinarine as a treatment for leukemia and basal cell carcinomas. However, extreme caution is needed in using these compounds because of their toxicity, which is not often labeled on herbal remedies.

Chelone glabra L., white turtlehead (Plantaginaceae)

Turtlehead is appropriately named because the shape of the flower resembles the head of a turtle. Linnaeus gave the name to the genus based on a story in Greek mythology, where the nymph Chelone was cast into a river, along with her house, for insulting the gods.

The setting is appropriate because turtlehead, like its namesake, is associated with water. It favors full to partial sun in ditches, rich fens, and margins of streams and lakes. Two to

three feet tall and arising from a rhizome system, it has opposite, toothed leaves. Flowers are white, sometimes with a tinge
of pink, and borne in rows in mid- to late summer. A cluster of
plants in full flower is striking. Widespread throughout eastern North America, turtlehead is state-protected in New York.

Veronica americana Schwein. ex Benth.,
American brooklime (Plantaginaceae)

A true aquatic in some situations, American brooklime can
grow submerged with floating leaves. More frequently, it is
found in marshes, seeps, and along streambanks. The intrigu-
ing common name "brooklime" is apparently a corruption of
an old word for mud, appropriate for the habitat of this species.

Stems are weak, often bending and rooting. Leaves are opposite (except those associated with the flowers) and are sharply toothed and narrowly oval in shape. Flowers are terminal, light blue to white, and last for only a day. Fruits are round and flattened.

This and other members of the genus are usually called "speedwell," based on an old greeting, "speed well," much as we might say, "Godspeed." The name was applied to species of *Veronica* because the corollas fall off within a day—speedily.

Veronica officinalis L., common speedwell (Plantaginaceae)

This is the "official" speedwell, the term *officinalis* being derived from the same root as "office" and meaning "authorized" or "approved." In this case, common speedwell is approved for its medical use. Numerous cures are attributed to it in homeopathic medicine.

It is a common weed throughout the United States as well as in Europe, where it originated. Plants are small, seldom more than ten inches tall, with opposite, elliptical, and toothed leaves. Flowers are on stalks growing out of the leaf axils and are colored blue, lavender, or white. Like all members of the genus, there are only two stamens.

There are at least six other species of *Veronica* in the Adirondacks, and likely more will spread there. Of these, two—*Veronica scutellata* L., marsh speedwell; and *Veronica serpyllifolia* L., common thyme-leaved speedwell—are native.

Persicaria sagittata (L.) H. Gross, arrow-leaved
tearthumb (Polygonaceae)

This scrambling vine of wetlands will tear much more than a
thumb; the narrow stems are covered with recurved prickles
that lacerate flesh. Human harm is, of course, not the chief role
of the armament. Rather, it is a means of distribution, as the
plant catches on animals, which then transport it and its seeds.
The sharp outgrowths also discourage grazing by insects and
other animals.

Leaves are alternate and arrow-shaped (sagittate) and char-
acterized by a papery tube around the stem at the base of the
leaf called an "ocrea." Flowers are small, with five or more white
or pink sepals, and are produced on short stalks in the axils of
the leaves. These small heads contain five to ten flowers per
cluster. Fruits are one-seeded, black, and trigonous (triangular

in cross-section). They are an important food for waterfowl.

The halberd-leaved tearthumb, *Persicaria arifolia* (L.) Haraldson, also occurs in the Adirondacks but is infrequent. It is coarser than arrow-leaved tearthumb and has larger, ax-shaped leaves. Like its smaller cousin, it can also inflict pain.

Lysimachia borealis (Raf.) U. Manns & Anderb., starflower (Primulaceae)

Starflower is a signature plant of Adirondack forests. It is also known by the older name of *Trientalis borealis*. From a narrow rhizome a wiry stem grows to about six inches, crowned with a whorl of four to eight or more leaves. The number and size of the leaves are variable. A typical plant has five lance-shaped leaves and one or two additional leaves that are much smaller. Arising from the middle of the terminal cluster of leaves is a single flower (but sometimes as many as four) with five or more white petals. The blossom is star-shaped, and the crown of leaves is often star-shaped as well. A few seeded capsules are produced in late summer.

This is a shade-tolerant plant often found growing with Canada mayflower (*Maianthemum canadense*) and bunchberry (*Cornus canadensis*). It is often abundant in peaty soil and decomposing woody debris, as well as rocky, acidic habitats.

Actaea rubra (Aiton) Willd.,
red baneberry (Ranunculaceae)

The common name is descriptive—the fruit is the bane of anyone who ingests it, owing to the toxicity of the bright red berries (in some areas white-fruited plants are frequent). Red baneberry is common in the Adirondacks in a variety of wooded

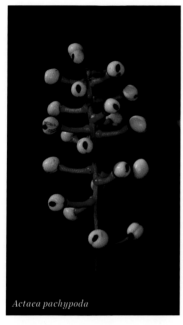

Actaea pachypoda

habitats. Plants are up to four and a half feet tall, with usually three, large doubly compound leaves, that is, each leaflet is further divided into several more leaflets. White flowers are produced in the spring on a spike at the tip of the plant. Flowers have a variable number of petals, four or more, and long stamens exceeding the petals. Fruits are shiny red and mature in mid- to late summer.

Equally widespread is the white baneberry, *Actaea pachypoda* Elliott, similar in both habit and habitat to red baneberry. As the common name implies, it has white fruits with a black spot at the tip, giving it another common name, "doll's eyes." Confusingly, the white baneberry can sometimes have red fruits. How to tell them apart? Although there are subtle differences in the petals, the surest way to separate the two is by the color and shape of the fruit stalk (the pedicel). White baneberry has a thick, red pedicel that is thicker than the main fruit axis, while the pedicel of red baneberry is thinner and brown.

Both species are being studied for medicinal uses, like their well-known and widely utilized relative black cohosh (*Actaea racemosa* L.).

Anemone canadensis L.,
Canada anemone (Ranunculaceae)

Anemone virginiana

The genus *Anemone* is named after the wind that blew the petals from the flower that sprang from the blood of the Greek god Adonis when he was murdered. That is why some species of anemone are called "windflowers," because they appear during the windy month of March.

Canada anemone flowers in the spring. It is a tall, single-stemmed plant with alternate, deeply divided, stalked stem leaves with three to five lobes that have pointed tips. Just below the flower is a set of smaller unstalked leaves (bracts) that subtend the large, white flower. Damp thickets, margins of damp woods, and fens are habitats for Canada anemone.

Tall anemone, *Anemone virginiana* L., also occurs in the Adirondacks. Overall, it is a taller plant with branched stems and floral bracts that are stalked and flower petals that are notched. The striking difference, however, is the fruit of tall anemone, which forms a thimble-shaped cluster of fruits yielding densely woolly seeds.

Anemone quinquefolia L.,
wood anemone (Ranunculaceae)

Large patches of this attractive, white-flowered, diminutive spring ephemeral are produced by an extensive rhizome system from which the stems, seldom more than six inches tall, arise. Leaves are alternate and divided into three leaflets. Like other members of the genus, wood anemone has a series of bracts (modified, reduced leaves) at the base of the flower. Petals (actually sepals) are five or more. Flowers are white (occasionally pink), with a green center. Moderately moist upland forests are its chief haunt, though it can also be found in floodplain forests. It is common throughout the Adirondacks.

Clematis virginiana L., virgin's bower (Ranunculaceae)

Often clambering over supporting vegetation, virgin's bower is the most common native clematis in the Adirondacks.
It occurs in wet meadows, woodland borders, and along roadsides and streambanks. This semi-woody climber has opposite,
three-parted leaves with a long tip, the only climbing plant in
the flora with this combination of characters. It is dioecious (*di*,
"two"; *oikos*, "home"), meaning each plant has flowers of only
one sex—the only clematis in the flora to have unisexual plants.

Virgin's bower produces prodigious numbers of fragrant,
white, unisexual flowers in mid- to late summer. Both male
and female flowers have four petal-like structures, sometimes
tinged with purple. Every female flower produces about twenty-five fruits, each dry and one-seeded, with a long, feathery
extension. Fruits are carried by animals.

The introduced aggressive relative, yam-leaved clematis (*Clematis ternifolia* DC.) resembles virgin's bower but has bisexual (perfect) flowers; thicker, elliptical leaves without an extended tip; and is usually restricted to disturbed sites. It is not common in the Adirondacks but has been expanding its range.

Coptis trifolia (L.) Salisb., goldthread (Ranunculaceae)

This low, modest plant produces a three-lobed evergreen leaf that arises from a creeping rhizome. Flowers are specialized. The white "petals" are actually sepals, with the petals modified into tiny, spoon-shaped nectar-producing structures that play a role in insect pollination. The fruits, produced in midsummer, look like an umbrella with the fabric gone and only the staves remaining. The "staves" are minute follicles with tiny seeds.

An iconic plant of the northern forest, goldthread was once collected by the wagonload for the production of bitters and was a major part of the Shakers' herbal medicine business. Bitters were popular in the Victorian era and thought to be salubrious before a meal, taken as a kind of tonic. Goldthread was also a sought-after cure for canker sores, so "canker root" is another common name of this evergreen creeping denizen of Adirondack mossy woods and bogs.

Goldthread is trans-boreal in its distribution, occurring all across North America. Other species of *Coptis* are used as bitters and medicines in China and India. The common name is derived from the golden, wiry rhizome, whose distinctive color results from a high concentration of the bitter compound berberine.

Hepatica americana (DC.) Ker Gawl., round-lobed hepatica (Ranunculaceae)

This is one of the best known of all spring wildflowers and is broadly distributed in the Adirondacks. The widely used common name, "hepatica," is a classic example of the Doctrine of Signatures. This ancient teaching states that if a plant part looks like a body part, that plant or plant part will affect the corresponding organ of the human body. The evergreen leaves of hepatica are tri-lobed with rounded tips, suggesting a fanciful resemblance to the human liver and therefore having purported efficacy in treating liver ailments.

For most of us, however, it is not the medicinal value of the plant that is our chief interest but rather the attractive flowers and leaves. Growing close to the ground, hepatica is only a few inches high and in the early spring produces white or blue flowers.

A related and similar species, *Hepatica acutiloba* DC., is distinguished from *Hepatica americana* by its acute leaf tips. There is also a difference in habitat, with *H. acutiloba* favoring richer soil. Flower color in this latter species is also quite variable, exhibiting various shades of white, purple,

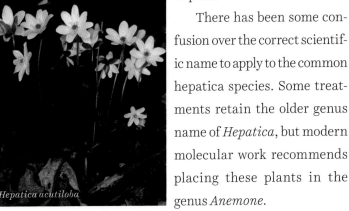

Hepatica acutiloba

or pink.

There has been some confusion over the correct scientific name to apply to the common hepatica species. Some treatments retain the older genus name of *Hepatica*, but modern molecular work recommends placing these plants in the genus *Anemone*.

Thalictrum pubescens (L.) Pursh,
tall meadow rue (Ranunculaceae)

This stately plant is one of our tallest wildflowers. A hardy perennial growing up to six feet, meadow rue often forms clumps with several stems and prefers open, sunny areas like roadsides and margins of lakes and rivers. Meadow rue plants are unisexual, that is, there are male plants and female plants. Because this species lacks petals, the male plants are more conspicuous when blooming in midsummer with their masses of small white flowers. Contrariwise, the female flowers are green and easily overlooked. By late summer, the small, single-seeded, beaked fruits are produced.

The common name comes from the superficial appearance of the leaves to leaves of the rue plant (*Ruta graveolens* L.), a medicinal plant more familiar to the Europeans who named the genus than it is to most Americans. The compound leaves with their numerous small leaflets give the plants a fernlike appearance.

Fragaria virginiana Mill.,
common wild strawberry (Rosaceae)

Everyone knows strawberries, one of the most popular fruits in the United States. Common wild strawberry looks like the cultivated strawberry, but the fruit is generally smaller and, in the opinion of many, tastier. In fact, this common Adirondack plant of clearings, meadows, pastures, and roadsides is one of the parents of the commercial strawberry, *Fragaria* × *ananassa* (the multiplication sign is botanical shorthand for a hybrid), chosen because of the sweet flavor. The other parent is *Fragaria chiloensis* (L.) Mill. from Chile, selected because of the size of the fruit.

Common wild strawberry is a small plant, seldom more than a few inches tall, with leaves that are divided into three leaflets. Flowers are white with five petals and vary in their sexual permutations. Some flowers are male, others female, and others have both sexes. Strawberry fruits are familiar, but their structure is unusual. Technically, the fruits are the tiny, strawlike seeds (botanically called "achenes"), borne in small depressions on the swollen, red tip of the flower stem. After fruiting, the plant sends out runners that grow along the ground and root, producing a patch of strawberries.

Another species of strawberry occurs in the Adirondacks— *Fragaria vesca* L., which comprises two subspecies of the woodland strawberry, one from Europe, the other from the United States. This is the most wide-ranging *Fragaria*, distributed across North America, Eurasia, and North Africa. The two species look superficially alike, but the woodland strawberry has achenes produced on the surface of the fleshy receptacle, not in small pits. Likewise, the teeth of the leaves of this species are uniform; leaves of the common wild strawberry have two different kinds of teeth at the leaf tip.

Rubus repens (L.) Kuntze, dewdrop (Rosaceae)

Studying this creeping member of the rose family close-up requires the practice of belly botany. You have to get close to the ground to see it. Dewdrop occurs in mixed pine and hardwood forests in small patches. It has simple leaves about two inches long, with prominent teeth and veins that radiate from the heart-shaped base of the leaf. The plant is also known as "false violet" because of the similarity of its leaf to those of many species of violets. Among the differences between true violet and false violet are the very hairy leaf stalks of false violet.

Two kinds of flowers are produced. The most apparent are the conspicuous white flowers with five petals. Remarkably, these seldom produce seed; it is the small, closed flowers that do so. These cleistogamous (non-opening) flowers can be located by removing leaf litter around the plant and finding the tiny buds borne on a sharply recurved stalk. The fruit is technically edible, with a taste that with some imagination resembles that of a raspberry. However, it would take at least ten fruits of the dewdrop to equal the size of a single raspberry.

Recent research has placed dewdrop in the genus *Rubus*, a group that includes blackberries and raspberries, replacing the more familiar name *Dalibarda repens* L.

Sibbaldia tridentata (Aiton) Paule & Soják, three-toothed cinquefoil (Rosaceae)

Rocky ridges, alpine meadows, and rocky outcrops are the habitat of this cinquefoil, so named because most species have a five-parted leaf—*cinquefoil* in French. It appears to be an herbaceous plant but is technically an evergreen creeping subshrub. Like many alpine tundra plants, its leaves are borne close to the ground. Leaves are three-parted, shiny, thick, and leathery, often turning red in winter. Flowers are borne in mid- to late summer and have five white petals.

Three-toothed cinquefoil is part of a guild of plants of the most exclusive community in the Adirondacks—the High Peaks. Fewer than one hundred acres of this fragile plant community exist in the Adirondacks.

Mitchella repens L., partridge berry (Rubiaceae)

The coffee family, Rubiaceae, is one of the five largest plant families, largely tropical in distribution and a small component of the Adirondack flora, with only five genera. The only evergreen member of the family here is partridge berry, a common plant in a diversity of habitats throughout the region. The genus is named after a colonial Virginia physician, John Mitchell (1711–68), best known for his map-making skills.

This is a winsome, miniature vine that can form large colonies through its ability to root and branch. Attractive white flowers appear in early summer. Superficially, all the flowers look alike, but on careful examination two types are evident. Some plants produce flowers with exserted (extending beyond the floral tube) styles, others with inserted (within the floral tube) stamens, an adaptation that ensures cross-pollination. Pollen from long-style plants will only fertilize short-style plants, and vice versa.

The attractive, bright red fruits are unusual in being formed by the fusion of the ovaries of two separate flowers. Fruits are edible but flavorless—more of a nibble than a food.

Comandra umbellata (L.) Nutt.,
bastard toad flax (Santalaceae)

This wide-ranging species, found in forty-seven states and the majority of Canadian provinces, usually occurs in open, sunny areas. Seldom more than a foot tall, with small, alternate leaves, bastard toad flax is strongly rhizomatous and can form large patches. Small, white, five-parted flowers appear in the spring. The fruit is semi-fleshy, purplish, and one-seeded but seldom seen.

The presence of chlorophyll belies the parasitic nature of this plant, which connects to its host via modified roots termed "haustoria." The numerous small haustoria are readily seen when the plant is dug and the roots of surrounding host plants are examined. The host range is very broad but little is known about any host preference. As if one level of parasitism were not enough, this unassuming plant is the alternate host of pine rust (*Cronartium comandrae* Peck), a fungus that can infect and cause damage to hard pines.

The varying levels of parasitism are not the reason for the common name, bastard toad flax. Rather, it is based on the suggestion that the plant looks like toad flax (species of *Linaria*).

Micranthes virginiensis (Michx.) Small, early saxifrage (Saxifragaceae)

This perennial plant of seeps is usually found on hillsides in full sun, but it also grows on large rocks in hardwood forests. As the common name suggests, it flowers in early spring. All the leaves are basal and oval-shaped. The flowering stem is covered with sticky hairs that may impede insects. Flowers are borne on spikes up to about sixteen inches tall. They are white or yellowish, small, and five-parted with ten yellow stamens.

Mitella diphylla L.,
two-leaved mitrewort (Saxifragaceae)

Mitella nuda

The two-leaved mitrewort is one of the most distinctive spring wildflowers, growing in moderately moist, rocky forests and seepage areas, often on base-rich (high pH) substrates. A single pair of opposite, three-lobed leaves arise at mid-stem, below a simple flowering stem. Other leaves are clustered at the base of the stem of this rhizomatous plant. The flowers are unique among Adirondack flora in having five white petals that are fimbriate (divided into very narrow strips). The fruit is a shallow, cup-shaped structure loaded with tiny, shiny black seeds that are distributed by raindrops.

A second species of the genus is represented by the smaller-statured naked mitrewort (or naked bishop's cap), *Mitella nuda* L., often found on mossy substrates. "Naked" applies to the appearance of the stem, as there are no mid-stem leaves in naked mitrewort. It has greenish-yellow instead of white petals, and slightly larger but many fewer flowers than the previous species.

Tiarella cordifolia L., foamflower (Saxifragaceae)

Foamflower is an attractive, clump-forming perennial with heart-shaped (*cordifolia* means "heart-shaped") leaves at the base of the plant. The leaf stalks (petioles) are often covered with glandular hairs. As in most members of the saxifrage family (Saxifragaceae), the leaves have a palmate pattern of venation.

Masses of white flowers appear in early spring. The foamy appearance results from the numerous stamens that extend considerably beyond the throat of the five-parted flower. It has narrow petals, so foamflower is sometimes confused with mitrewort. Foamflower grows most frequently at lower elevations on moderately moist forest slopes, seepages, and swamps.

Because of its attractive flowers and foliage, foamflower is frequently used in gardens, especially in shaded settings. Several cultivars with pinkish flowers and red leaf venation are available.

Laportea canadensis (L.) Wedd., wood nettle (Urticaceae)

An irritating companion of many of the best-known spring ephemerals, wood nettle grows up to four feet tall and is covered with the same stinging hairs that define the weedy stinging nettle, *Urtica dioica*. The two are readily separated, however, by the leaf arrangement—opposite in stinging nettle, alternate in wood nettle (there may be some opposite leaves but most are alternate). Also, wood nettle is a native plant of moderately moist to wet forests in the Adirondacks, while stinging nettle is introduced and most common in sunny, disturbed sites.

Flowers of wood nettle are unisexual, small, and inconspicuous. Male flowers are white or whitish-green and produced on
side branches, while female flowers are green and terminal.

Like stinging nettle, the young leaves and shoots of wood
nettle are edible after boiling.

Viola canadensis L., Canada violet (Violaceae)

Viola pallens

Canada violet grows in rich, moderately moist forests and is
widespread in the Adirondacks. This stemmed violet is distinguished by white flowers with a yellow center. It flowers in midspring and is often cultivated in gardens. Unlike some other
violets, it does not spread via runners but often forms large,
attractive clumps. It is one of the most wide-ranging North
American violets.

Three additional white-flowering, but stemless, violets in
the Adirondacks are *Viola blanda* Willd., sweet white violet;
Viola pallens (Banks ex Ging.) Brainerd, smooth white violet;
and *Viola renifolia* A. Gray, kidney-leaved violet (Violaceae). Of
these, the kidney-leaved violet has the most distinctive leaves,
as the common name suggests. This species is found in moist

to wet sites, including evergreen forests. The sweet white and smooth white violets have the more typical heart-shaped or round leaves, but the smooth white violet (listed as *V. maclo-skeyi* Lloyd in many guides) has smaller leaves than the sweet white violet and is generally found in wetter sites, including along streams.

Green Flowers

Arisaema triphyllum (L.) Schott,
common Jack-in-the-pulpit (Araceae)

The imagery inherent in the common name of this well-known spring wildflower tends to be lost on present-day students, because few have seen a traditional pulpit as found in cathedrals and old churches. In what way does the plant looks like a pulpit? The flowers are on a rod-shaped structure, the spadix, surrounded by a wrapping called the spathe. "Jack," the preacher/spadix, stands erect inside this sheathing, and his "head" extends beyond the lip of the spathe as in an old-fashioned pulpit.

Each plant characteristically has a single, three-parted leaf. Jack-in-the-pulpit flowers in the spring along with its coterie of spring ephemerals. The small flowers on the spadix are greatly reduced and unisexual. Insects are drawn into the spathe by volatiles produced by the plant. The flowers may be simple, but the life cycle of this widespread and often abundant Adirondack plant is complex. First of all, it can change sex. After a heavy fruiting year (from a plant obviously in the female stage) the plant often reverts to the male stage.

There are also taxonomic complexities. Several subspecies are recognized, and each of these has a different number of chromosomes. *Arisaema triphyllum* (L.) Schott var. *stewardsonii* (Britton) Huttl., swamp Jack-in-the-pulpit, is easy

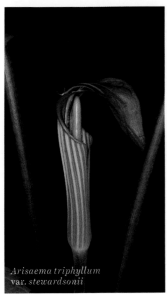

Arisaema triphyllum var. stewardsonii

to recognize with its white-and-green-striped spathe in its characteristic habitat.

Fruits are produced in the fall, a cluster of bright red fleshy fruits (berries) with one to six seeds in each berry. The entire plant, especially the corm, contains raphides, needle-like crystals of calcium oxalate that, when ingested, puncture mucous membranes. In severe cases, the resultant swelling of airways

results in asphyxiation. The fleshy, starch-filled corm is sometimes eaten. Boiling destroys the raphides, and the oxalate is
water soluble.

Aralia racemosa L., spikenard (Araliaceae)

This species has perhaps the largest leaves of any Adirondack
plant—often more than two feet across, arising from the base
of the usually multi-branched plant. Each leaf is divided into
leaflets that are again divided into leaflets. As a result, a single plant is wider than it is tall. It prefers partial shade and
grows on rocky slopes and ledges. Greenish-white flowers are
produced in umbels (globe-shaped clusters) on elongated spikes
in midsummer. The fruits have been likened to those of the
black elderberry, *Sambucus canadensis*, in appearance but not
in taste.

True spikenard, as used in ancient times and mentioned in the Bible, is from an unrelated Himalayan plant, *Nardostachys jatamansi* DC. The Adirondack species is called spikenard because its fragrance is purportedly similar to that of the biblical spikenard.

Orthilia secunda (L.) House, one-sided wintergreen (Ericaceae)

One-sided wintergreen, also commonly referred to as "one-sided shinleaf," occurs in dry to moist forests, often with eastern white pine (*Pinus strobus*). Six to twenty white to greenish, five-petaled flowers less than one-half of an inch in diameter are packed along only one side of the unbranched, curved flower stalk that is up to eight inches tall. Flowering occurs from early to midsummer. The evergreen leaves are elliptical to roundish, and in a whorl-like cluster. Plants typically occur in patches.

Veratrum viride Aiton, false hellebore (Melanthiaceae)

One of several very toxic plants in the Adirondack flora, false hellebore is also one of the most widespread. The origin of the common name, false hellebore, is unclear, as the plant bears little resemblance to true hellebore, a species of *Helleborus* (Lenten rose) that has compound leaves and large flowers. False hellebore is a large, coarse, perennial plant up to six feet tall with alternate leaves that encircle the stem. Leaves are broad—up to five inches wide, with very deep, conspicuous veins. Small, greenish-white flowers are produced in large numbers in the spring. The fruit is a capsule with a few seeds. This long-lived perennial develops a large, knobby rhizome from which the individual stems arise. Large stands of false hellebore can be found along streams, in seepage areas in forests, and in marshes. It takes several years for the plant to reach maturity and flower.

Like many toxic plants, false hellebore has been employed as a medicine and is still used by practitioners of homeopathy and other plant-based remedies. Extreme caution is required in its use, however, because the various powerful alkaloids present throughout the plant (and concentrated in the rhizome) are dangerous and can cause death.

Corallorhiza trifida Chat., early coralroot (EV) (Orchidaceae)

Early coralroot, also commonly listed as the "early green coralroot," blooms in mid- to late spring, about one month earlier than spotted coralroot, and occurs in conifer swamps, often associated with northern white-cedar (*Thuja occidentalis*). Its stems, as well as flowers and fruit capsules, have a small amount of chlorophyll and hence are yellowish-green. The plant does carry out photosynthesis and still requires fungi to meet its nitrogen and carbon needs. Its flower has a white lip, with no spots.

Early coralroot more commonly occurs in larger clumps and larger colonies than spotted coralroot, with over a dozen stems in clumps and hundreds of stems in colonies. The showy lady's slipper (*Cypripedium reginae*) is often associated with early coralroot, although the former blooms about one month later.

Liparis loeselii (L.) Rich.,
fen twayblade (EV) (Orchidaceae)

Because of its short stature and greenish-yellow, small flowers, the fen twayblade often goes unnoticed while flowering in late spring to early summer, even though it is frequent in moist substrates that are not acidic. It is especially common in fens and in moist, disturbed places like sandy or gravelly substrates of meadows, borrow pits, ditches, and shorelines of lakes, usually lacking a dense forest canopy. In the best conditions, one can locate many dozens at a site.

The species is named for the German botanist Johann Lösel (1607–57). *Liparis* comes from *liparos* ("greasy" or "shining"), referring to its two ("tway") glossy and smooth basal leaves that are elliptical in shape. It is often easier to find when its fruit is maturing in a cluster of two to fifteen ellipsoid and yellowish capsules along the flowering stalk that is held above its leaves. This species is self-fertile, and raindrops are known to facilitate pollination.

Malaxis monophyllos var. *brachypoda* (A. Gray) F. Morris and E.A. Eames, white adder's mouth (EV) (Orchidaceae)

Malaxis unifolia Michx., green adder's mouth (EV) (Orchidaceae)

The two *Malaxis* (adder's mouth) species in the Adirondacks are similar in that both have single leaves, as their species names suggest (Greek, *monos phyllon*, "solitary leaf"; Latin, *unus folium*, "one leaf"). *Malaxis* is from the Latin *malacos* ("delicate"), a descriptive name for these delicate orchids. The flower stalks of both species bear up to fifty minute flowers. Flower stalks arise from an annual pseudobulb (a type of storage organ) that basically lacks roots and rhizomes, and produces a pseudobulb for the following year. Both species bloom in summer, with the white adder's mouth beginning a week or two earlier and green adder's mouth still blooming a week or two after the white adder's mouth is finished flowering.

Both species can occur together in rich, mossy, lowland conifer forests, but within these locations, the green adder's mouth tends to inhabit drier microsites like rotten logs and tip-up mounds. This latter species can also occur in drier, open woodlands and in conifer plantations; look for it too along roadsides and on ditch berms that are not densely vegetated. An abundance of black ash (*Fraxinus nigra*) can be associated with the white adder's mouth. These adder's mouths can be distinguished by the width of the inflorescence, or flowering stalk, which is relatively wide and cylindrical in shape, like a bottlebrush in appearance, for the green adder's mouth, versus a slender spike of flowers for the white adder's mouth. Individual flowers are also more noticeably stalked in the green adder's mouth and are yellowish-green in color, whereas the flowers of white adder's mouth are held closer to the stalk and are usually yellowish-white or pale white.

Neottia auriculata (Wiegand) Szlach., auricled twayblade (S1, EV) (Orchidaceae)

Neottia convallarioides (Sw.) Nutt. ex Ell., broad-lipped twayblade (S1, EV) (Orchidaceae)

The broad-lipped twayblade is one of the rarest orchids in New York State, occurring in deeply shaded conifer swamps, sometimes near streams. When found, it can be in colonies of hundreds of individuals. Although up to twenty flowers can be held on a single stalk in late spring to early summer, their small size and yellowish-green color are rather inconspicuous in this species' typically dark settings.

Neottia convallarioides

The auricled twayblade is another very rare twayblade in the Adirondacks. It is found in alder thickets, as well as under conifers along streambanks, and is the only *Neottia* in the Adirondacks with a light, blue-green flower color. Additionally, the lip of *N. convallarioides* is stalked and broader than the sessile lip of *N. auriculata*.

The genus of *Listera* has recently been changed to *Neottia* for all of the species in New York. Of all the orchids in the Adirondacks, these *Neottia* species are the ones for which new locations are most likely to be discovered, as more people explore their habitats and learn to recognize them. Search for them on a sunny day during their flowering period, and hope that they are in a patch of light within their otherwise exceptionally shady surroundings.

Platanthera aquilonis Sheviak, northern green bog orchid (EV) (Orchidaceae)	*Platanthera huronensis* (Nutt.) Lindl., Lake Huron bog orchid (EV) (Orchidaceae)

Until rather recently, *P. aquilonis* was known primarily as *Platanthera hyperborea* (including *P. hyperborea* var. *hyperborea*). The latter species is now believed to be restricted to Iceland. *Platanthera huronensis* is similar to the true *P.*

hyperborea, although all three species are difficult to separate. *Aquilonis* is from Latin, meaning "northern," and refers to this species' northern distribution in North America; *huronensis* refers to Lake Huron.

The northern green bog orchid and Lake Huron bog orchid both occur in fens, wet ditches, and at stream edges, typically in wet substrates influenced by calcium, and usually not in true bogs. Both can be rather robust, reaching nearly thirty inches in height, although those individuals that are nearly forty inches tall are usually *P. huronensis*. These species often co-occur, and their yellowish-green flowers open at about the same time during the summer. Living, flowering specimens can be tediously distinguished by examining with a 10× hand lens the arrangement of the two anthers. In *P. aquilonis*, the anthers are widely spaced at the base, adjacent to the stigma, but converge above it, forming a hoodlike structure. In *P. huronensis*, the anthers are nearly parallel from bottom to top, not nearly touching above the stigma. While this difference seems like something only a botanist would notice, this characteristic determines which insects can pollinate either species. And for those with a keen sense of smell, *P. huronensis* has a more noticeable fragrance than *P. aquilonis*, and a whitish-green lip that is expanded at the base instead of the yellowish lip that is evenly tapered in *P. aquilonis*. Alternatively, even the most serious wildflower enthusiasts can accept that these two species are incredibly similar and still enjoy them as semi-anonymous individual specimens.

Platanthera clavellata (Michx.) Luer,
little club-spur orchid (EV) (Orchidaceae)

The little club-spur orchid is similar to *P. obtusata* in its small stature but occurs in open and often disturbed areas with acidic, moist, sandy, or peaty substrates, including margins of lakes, roadside ditches, meadows, and occasionally bogs (not deeply shaded conifer swamps dominated by northern white-cedar). Both species usually have a single basal leaf, with smaller, bractlike leaves often pressed against the stem. The little club-spur orchid blooms during midsummer, and its flowers, with a clublike (*clavellata*) thickening at the tip of the spur, are distinctly twisted along the flowering stalk. This is one of the most frequently encountered orchids in the Adirondacks; one can expect to find dozens of scattered plants at a single location.

Platanthera flava (L.) Lindl. var. *herbiola* (R. Br.) Luer, northern tubercled orchid (EV) (Orchidaceae)

Because of its dark green leaves and stems and its yellow-green flowers, the northern tubercled orchid is often not readily seen in the dark conifer swamps in which it occurs. Plants flower in early to midsummer and are usually over one foot tall. This orchid also occurs in floodplains, other wet woods, and lakeshores. Occasionally, numerous flowering stems can be found in small but dense patches.

Flava ("yellow") refers to its flower color. The common name refers to the tubercle, or small swelling or knob, which is round and on the upper lip of the flower. This bump apparently forces pollinators to enter on either side, thus facilitating more effective pollination. The other variety of this species, *P. flava* var. *flava*, occurs in the southeastern United States.

Platanthera hookeri (Torrey ex A. Gray) Lindl., Hooker's orchid (S1, EV) (Orchidaceae)

Named after Sir William Jackson Hooker (1785–1865), professor of botany and director of the Royal Botanic Gardens at Kew in the mid-1800s, this orchid occurs in rich, moist, upland

forests dominated by sugar maple and American beech, blooming in early to midsummer. It can also be found in drier forests
with a mixture of hardwoods and conifers. Its pair of widely
elliptical to nearly round basal leaves lying close to the ground
resembles those of *P. macrophylla* and *P. orbiculata*, which also
occur in these habitats, but the flowers of Hooker's orchid are
greenish to yellowish rather than whitish. Without flowers, the
leaves of Hooker's orchid are not as glossy, nor as round or large.
Although this species is not named after the distinctly hooked
lip, that feature also separates Hooker's orchid from these other two species.

Platanthera lacera (Michx.) G. Don, ragged fringed orchid (EV) (Orchidaceae)

The ragged fringed orchid
is one of the more com-
mon *Platanthera*s in the
Adirondacks, occurring in
open wet meadows, fens, roadsides, and bogs—it is one of the
least habitat-specific of New
York orchids, as long as the
site is sunny. Despite the "ragged" in its common name, it
is sometimes among the more
beautiful fringed orchid species, with its many, highly

dissected greenish-yellow to greenish-white flowers per stalk
in late spring into summer. Individual plants can reach nearly
thirty inches in height and often occurs in pairs. *Lacera* refers
to the lacerated or torn appearance of the lip.

Platanthera obtusata (Banks ex Pursh) Lindl.,
blunt-leaved orchid (EV) (Orchidaceae)

The blunt-leaved orchid is found in conifer swamps of northern
white-cedar, black spruce, white spruce, balsam fir, and tamarack. The best time to look for it—unfortunately—is during

the peak of mosquito season. Its green to greenish-white flowers open in early to midsummer and are pollinated by mosquitoes and small moths. *Obtusata*, meaning "blunt," refers to the rounded apex of the single leaf on the short (less than six inches tall) plant.

Rumex acetosella L., sheep sorrel (Polygonaceae)

Sheep sorrel was introduced from Eurasia and is widespread in North America, occurring in lawns, fields, rocky ledges, and a diversity of other open, disturbed sites in all fifty states. Aggressively spreading from thin rhizomes, sheep sorrel at maturity is less than two feet tall, with leaves having a narrow, triangular shape like that of a halberd spear. It blooms throughout the growing season; the inconspicuous flowers, arranged in small whorls, are greenish but turn red with age. Individual plants vary in terms of male and female flowers; some are unisexual, while others produce flowers of both sexes. The fruit is three-sided, dry, and hard.

Sheep sorrel is a wild edible that received its common name because its pleasantly sour taste is similar to that of several other plants also known as "sorrel." Animal names were often employed to describe plants that were considered not as good as others, for example, hog peanut and horse chestnut. Large quantities, however, should not be ingested because of the presence of oxalic acid, which, in some individuals, can contribute to kidney stones.

Sheep sorrel should not be confused with common sorrel, *Rumex acetosa* L. Though the scientific names sound similar, the two species are taxonomically distinct. Common sorrel is much less frequent than sheep sorrel and, like that species, is an introduction from Eurasia. The broad-leaved or bitter dock, *Rumex obtusifolius* L., also occurs in the Adirondacks on similar, disturbed sites. It is a much larger plant than sheep sorrel, with large, oblong leaves having very wavy margins.

Potamogeton natans L., floating-leaved pondweed (Potamogetonaceae)

Pondweeds, species of the genus *Potamogeton*, are the largest group of truly aquatic plants in the Adirondacks, where ponds, streams, flowages, and lakes provide diverse habitats. All pondweeds are obligate aquatic plants: they must grow in water. At least seven species are frequent in the region and several more occur sporadically. Taxonomically, they can be difficult to distinguish; identification relies on characteristics of leaf venation, fruit shape, and rhizome color.

All are perennial, rooted plants. Often forming large stands in full sun and in water five feet or less deep (but water depth can be quite variable for this species), the plants flower in midsummer. Even though they are highly specialized for their watery world, reproduction takes place out of the water. Those that have been studied are anemophilous (wind pollinated), and the manner in which the flowers are borne is an adaptation for anemophily. The flowers lack showy petals and are arranged on narrow spikes that extend above the surface of the water, where they can encounter air currents. The fruit is one-seeded, surrounded by fleshy tissue.

Floating-leaved pondweed is one of the most abundant pondweeds in Adirondack waters as well as across most of the United States, Canada, and Eurasia. It produces two kinds of leaves—floating and submersed. These differ strikingly in their appearance. The floating leaves are elliptical in shape; shiny, with a leathery texture; and are up to five inches long and two and a half inches wide with a rounded tip. They are borne on long leaf stalks. Submersed leaves are grasslike and lack differentiation into leaf stalk and blade. They are only a half-inch wide but up to six inches long. Like most aquatic plants, leaf size and shape is variable and influenced by water depth and light.

Floating-leaved pondweed, like other pondweeds, provides a valuable habitat for fish as well as food for aquatic animals and waterfowl. In parts of Asia and Africa, it has become a pest in irrigation and hydroelectric schemes. Young shoots and rhizomes are reported to be edible but hardly worth the effort of collecting in mud and water.

Sparganium americanum Nutt., American bur-reed (Typhaceae)

This is one of the most common aquatic plants in the Adirondacks, occurring in flowages and shallow water of lakes and ponds, often in mucky soil. Plants can have two habits. Those in deeper water have straplike leaves floating on the surface; those in stream margins and marshes have erect leaves. The seven species of bur-reed in the Adirondacks can all be found in similar habitats. Species are separated on mainly technical features, such as the number of stigmata and the shape and position of the sepals.

American bur-reed leaves can be distinguished from other wetland plants with similar leaves (cattails, irises) by the rounded tip of the leaf and its trigonous structure (triangular in cross-section). Flowers are unisexual and scattered along the upper part of stems and branches in dense, globose heads. Flowers have inconspicuous petals and sepals. The female heads, usually below the male flowers, develop into bur-like clusters of one-seeded fruits. These are valuable food for waterfowl. Bur-reed species have been used in bioremediation because of the plants' ability to sequester heavy metals and other toxins.

Typha latifolia L., *T. angustifolia* L., and *T.* ×*glauca* Godr., cattails (Typhaceae)

The most abundant and familiar marsh plants in the Adirondacks are the two species of cattails, wide-leaved cattail (*Typha latifolia*) and narrow-leaved cattail (*T. angustifolia*). They are superficially similar though easily separated

Typha ×glauca

upon close examination. The two species hybridize to form the hybrid cattail (also called "white cattail"), *T. ×glauca*.

All cattails have an extensive network of thick rhizomes. From these arise individual plants with their long, straplike leaves. Each flowering stem contains unisexual flowers, with the yellow male flowers above the green females. A single flowering stem has thousands of wind-pollinated flowers. One plant can produce up to seven hundred thousand plumed, one-seeded fruits in late autumn.

A simple criterion by which to separate species is the size of the gap between the terminal male flowers and the female flowers below. If the gap is noticeable—about the width of a thumb— it is the narrow-leaved cattail. If the male and female flowers are confluent, it is the wide-leaved cattail. A second feature is the height of the flowering stalks. Leaves are overtopped by the flowering stalk in wide-leafed cattail. Lastly, there is a difference in the width of the leaves, as the common names imply, though this is a less dependable feature.

It is likely that *T. angustifolia* is invasive, as its progression during the past few hundred years from brackish marshes on

the coast to inland waters is well-documented. Roadside populations are aided by heavy salt applications. The hybrid cattail
is also invasive. Both of these aggressive species are displacing
the wide-leaved cattail.

All parts of cattails are edible. In addition to being a valuable food source for wildlife, they are consumed by humans in
several parts of the world.

Urtica dioica L., European stinging nettle (Urticaceae)

Stinging nettle is an herbaceous perennial three to seven feet
tall. The soft, green leaves are opposite and one to six inches
long with a strongly serrated margin. Stinging nettle plants are
unisexual, that is, they have only male or only female flowers.
Flowers of both sexes are tiny, green, and inconspicuous. Seeds

are produced in late summer (on female plants, of course) and are edible.

The stinging hairs that cover the entire plant are its distinguishing feature. When touched, these hairs are transformed into needles that inject into the skin a cocktail of inflammatory chemicals such as acetylcholine, histamine, serotonin, and possibly formic acid, which cause a painful, localized irritation. Unlike the rash caused by poison ivy, this is not an immune response and does not spread to other parts of the body. It may appear that this introduction from Eurasia should be avoided. Au contraire. Stinging nettle is one of the tastiest of wild greens. Harvest when young—before flowering and while the stem is still tender—and boil. Boiling destroys the stinging hairs and their toxins.

References

Bramen, L. 2017. "From Ore to Orchids: The Legacy of Benson Mines."
Adirondack Life 48(1): 34–39, 69.

Bray, W. L. 1930. *The Development of the Vegetation of New York State*.
New York College of Forestry Technical Publication no. 29. 2nd
ed. Syracuse, NY: New York State College of Forestry at Syracuse
University.

Case, F. W., Jr., and R. B. Case. 1997. *Trilliums*. Portland, OR: Timber
Press.

Chapman, W. K. 1997. *Orchids of the Northeast: A Field Guide*.
Syracuse, NY: Syracuse University Press.

Edinger, G. J., D. J. Evans, S. Gebauer, T. G. Howard, D. M. Hunt, and
A. M. Olivero, eds. 2014. *Ecological Communities of New York
State*. 2nd ed. (revised and expanded edition of Carol Reschke's
Ecological Communities of New York State). Albany: New York
Natural Heritage Program, New York State Department of
Environmental Conservation.

Edsall, M. S. 1985. *Roadside Plants and Flowers*. Madison: University
of Wisconsin Press.

Heimburger, C. C. 1934. "Forest-Type Studies in the Adirondack Region." Ithaca, NY: Cornell University Agricultural Experiment Station, Memo 165, 1-122.

Henn, R. L. 2008. *Wildflowers of Ohio.* 2nd ed. Bloomington: Quarry Books (an imprint of Indiana University Press).

Homoya, M. A. 1993. *Orchids of Indiana.* Bloomington: Indiana University Press.

———.2012. *Wildflowers and Ferns of Indiana Forests: A Field Guide.* Bloomington: Indiana University Press.

Leopold, D. J. 2003. *Trees of New York: Native and Naturalized.* Syracuse, NY: Syracuse University Press.

———.2005. *Native Plants of the Northeast: A Guide for Gardening and Conservation.* Portland, OR: Timber Press.

Leopold, D. J., C. Reschke, and D. S. Smith. 1988. "Old-Growth Forests of Adirondack Park, New York." *Natural Areas Journal* 8:166–89.

Mathis, W. 2005. *The Gardener's Guide to Growing Hardy Perennial Orchids.* Doylestown, PA: Wild Orchid Company.

Raynal, D. J., and D. J. Leopold. 1999. *Landowner Guide to State-Protected Plants of Forests in New York State.* Syracuse, NY: State University of New York, College of Environmental Science and Forestry.

Reddoch, J. M., and A. H. Reddoch. 2007. "Population Dynamics and Flowering Synchrony of *Goodyera pubescens* (Orchidaceae) in Southwestern Quebec, Canada." *Journal of the Torrey Botanical Society* 134:379–88.

Slack, N. G., and A. W. Bell. 2007. *Adirondack Alpine Summits: An Ecological Field Guide.* Saranac Lake, NY: Adirondack Mountain Club.

Smith, W. R. 2012. *Native Orchids of Minnesota.* Minneapolis: University of Minnesota Press.

Sperduto, D., and B. Kimball. 2011. *The Nature of New Hampshire:
Natural Communities of the Granite State.* Durham: University
of New Hampshire Press.

Thompson, E. H., and E. R. Sorenson. 2000. *Wetland, Woodland,
and Wildland: A Guide to the Natural Communities of Vermont.*
Hanover, NH: University Press of New England, for the Nature
Conservancy and Vermont Department of Fish and Wildlife.

Tullock, J. 2005. *Growing Hardy Orchids.* Portland, OR: Timber
Press.

Weldy, T., D. Werier, and A. Nelson. 2018. *New York Flora Atlas.* [S. M.
Landry and K. N. Campbell (original application development),
USF Water Institute, University of South Florida.] Albany: New
York Flora Association. http://newyork.plantatlas.usf.edu.

Whiting, R. E., and P. M. Catling. 1986. *Orchids of Ontario.* Ottawa:
CanaColl Foundation.

Index to Species

About the Authors

DONALD J. LEOPOLD earned his PhD in forest ecology from Purdue University and his MSF in forest ecology and BS in ornamental horticulture / nursery management from the University of Kentucky. He is a Distinguished Teaching Professor at the State University of New York College of Environmental Science and Forestry, where he served as chair of the Department of Environmental and Forest Biology for thirteen years. He has taught numerous plant courses at the college's Cranberry Lake Biological Station in the western Adirondacks for over thirty years—instruction that provides the foundation for this book. He has also published nearly seventy journal papers, six books, and many additional pieces, all generally about topics in forest and wetland ecology, and native plants. His book projects generally incorporate his photographic images of plants taken during the past forty years. His research primarily focuses on the drivers of diversity and rarity at micro to macro scales; the restoration of unique plant communities on highly degraded sites; and applications of this research to sustainable urban landscapes and green solutions.

LYTTON JOHN MUSSELMAN is the Mary Payne Hogan Distinguished Professor of Botany at Old Dominion University, where he also served as chair of the Department of Biological Sciences. He received his BA from Beloit College, his MS from the University of Wisconsin–Milwaukee, and a PhD in botany from the University of North Carolina Chapel Hill. He has taught courses for more than four decades in the identification and use of plants at his home institution, as well as the State University of New York College of Environmental Science and Forestry at the Cranberry Lake Biological Station, the Au Sable Institute of Environmental Studies, and the University of Virginia Mountain Lake Biological Station. Recipient of four Fulbright awards, he has also been a visiting professor at the American University of Beirut; Aleppo University; the American University of Iraq, Sulaimani; and the Universiti Brunei Darussalam. Several of his books deal with plants of the Middle East. Closer to home, he is the coauthor of *Plants of the Chesapeake Bay: A Guide to Wildflowers, Grasses, Aquatic Vegetation, Trees, Shrubs, and Other Flora* and *The Quick Guide to Wild Edible Plants: Easy to Pick, Easy to Prepare.*